Code for Lighting

Society of
Light and Lighting

OXFORD AMSTERDAM BOSTON LONDON NEW YORK PARIS
SAN DIEGO SAN FRANCISCO SINGAPORE SYDNEY TOKYO

Butterworth-Heinemann
An imprint of Elsevier Science
Linacre House, Jordan Hill, Oxford OX2 8DP
225 Wildwood Avenue, Woburn MA 01801-2041

First published 2002

British Library Cataloguing in Publication Data
Code for lighting
1. Interior lighting 2. Interior lighting – Standards
I. Chartered Institution of Building Services Engineers
II. Society of Light and Lighting
729.2′8

Library of Congress Cataloguing in Publication Data
A catalogue record for this book is available from the Library of Congress

ISBN 0 7506 5637 9

For information on all Butterworth-Heinemann publications
visit our website at www.bh.com

Typeset by Keyword Typesetting Services Ltd

Printed and bound in Italy

Foreword

The *Code* has been published in various formats since 1936, this being the 16th edition. It is the standard reference on lighting design both within and outside the lighting profession. It is consistent with international (CIE) and European standards. This edition takes account of the 2001 revision of Part L of the UK Building Regulations (Part J of the Scottish Building Standards), as well as the forthcoming European standard on lighting the workplace.

In 1999 the Technical and Publications Committee of the Society of Light and Lighting (which has replaced the CIBSE Lighting Division) decided that, if the Society wished to keep its publications up to date and integrate them as a coherent set of guidance documents, a new approach was needed using electronic methods of publication. However, it was not obvious what would be the eventual medium and so the decision was made to publish the *Code for Lighting* on CD-ROM, but in a format which could in due course be transferred to another medium such as the Web.

This will enable the Society to update the *Code* annually, and to avoid having in circulation, in its various publications, similar but inconsistent information which has been published over a period of years. Eventually, the *Code for Lighting* will become a complete handbook of lighting recommendations covering all aspects of lighting, though it will be some years before this stage is reached.

Market research, however, indicated that although this approach would be popular with lighting professionals, it would not be convenient for the many outside the profession who make use of the Society's recommendations, and especially the *Code for Lighting*. The decision was therefore reached to publish a printed volume containing the three principal chapters of the *Code*: Visual effects of lighting; Recommendations (including the General Schedule); and Lighting design, along with the Glossary of terms.

Supervision of the revision was undertaken by the Technical and Publications Committee of the Society. Most of the actual work of turning the document into a form suitable for electronic publication, as well as updating to take account of European and international standards and UK legislation, was carried out for the Society by Peter Raynham of the Bartlett School of Architecture. The Society is grateful to the members of its Technical and Publications Committee and to Peter Raynham for making this new edition possible. Thanks are also due to Nick Paley of WSP Lighting for the design of the set of documents including the cover of this book.

Note from the publisher

This book is primarily intended to provide guidance to those responsible for the design, installation, commissioning, operation and maintenance of building services. It is not intended to be exhaustive or definitive, and it will be necessary for users of the guidance given to exercise their own professional judgement when deciding whether to abide by or depart from it. For this reason also, departure from the guidance contained in this publication should not necessarily be regarded as a departure from best practice.

An associated CD is available to purchase from CIBSE, and cross references to this CD are included in the text.

Acknowledgements

The cover image was designed by Neal Paley of WSP Lighting.

Photographs were kindly supplied by the following organisations: Art Ex Ltd, Concord Marlin, EDL, Megatron, Philips, Thorn, Zumtobel.

Related lighting publications available from CIBSE

Lighting Guide 1: The Industrial Environment (2002)
Lighting Guide 2: Hospitals and Healthcare Buildings (1989, addendum 1999)
Lighting Guide 3: The Visual Environment for Display Screen Use (2nd edition, 1996, addendum 2001)
Lighting Guide 4: Sports (1990, addendum 2000)
Lighting Guide 5: Lecture, Teaching and Conference Rooms (1991)
Lighting Guide 6: The Outdoor Environment (1992)
Lighting Guide 7: Lighting for Offices (1993)
Lighting Guide 8: Museums and Galleries (1994)
Lighting Guide 9: Lighting for Communal Residential Buildings (1997)
Lighting Guide 10: Daylighting and Window Design (1999)
Lighting Guide 11: Surface Reflectance and Colour – Its Specification and Measurement for Lighting Designers (2001)
Guide to Fibre-Optic and Remote-Source Lighting (2001, joint with Institution of Lighting Engineers)
Technical Memoranda 12: Emergency Lighting (1986, addendum 1999)
Technical Memoranda 14: Standard File Format for Transfer of Luminaire Photometric Data (1988)
Lighting the Environment: A Guide to Good Urban Lighting (1996)

Contents

Preface

Changes to the Code

This is the first edition of the *Code* published since the formation of the Society of Light and Lighting, but it follows the same format as the 1994 edition.

The principal changes to this edition of the *Code* have been driven by new and forthcoming standards and legislation, including a greater emphasis on energy efficiency.

The Schedule has been recast in a different format and follows in content tables 5.1 to 5.8 of the draft European standard prEN 12464 *Lighting of indoor work places*. It no longer includes sports premises since these are covered by their own European standard, BS EN 12193: 1999. Sports premises are dealt with in *CIBSE Lighting Guide 4*: *Sports* and its addendum.

The Unified Glare Rating system has replaced limiting glare index. This is again a result of publication of European standards, though UGR itself is described in a CIE publication rather than a European standard. It is expected that an increasing number of manufacturers will publish UGR tables for their luminaires. However, the numerical values differ little between LGI and UGR because of the way UGR is defined. Information is given to enable users of the *Code* to calculate UGR.

The section on energy and lighting has been recast both to tighten the energy consumption recommendations and to take into account publication of the 2001 edition of Part L2 of the Building Regulations, which covers non-domestic buildings (Part L1 covers domestic buildings but these are not, in general, within the scope of the *Code*). In Scotland the relevant, closely similar, but not identical, regulations are Part J of the Building Standards (Scotland) Regulations. Even now, the energy limits in the *Code* should be easy to achieve and there will often be scope for more efficient schemes than are required to meet the requirements of the Regulations.

Although the general format of the *Code* is unchanged, Part 3, dealing with lighting design, has been extensively recast to take account of changes in lighting practice and technology. The other Parts have been updated but their general structure remains unchanged.

Part 1

Visual effects of lighting

1.1 Introduction to the visual effects of lighting

The lighting of an interior should fulfil three functions. It should:

(*a*) Ensure the safety of people in the interior (Figure 1.1)

(*b*) Facilitate the performance of visual tasks (Figure 1.2)

(*c*) Aid the creation of an appropriate visual environment (Figure 1.3).

Safety is always important, but the emphasis given to task performance and the appearance of the interior will depend on the nature of the interior. For example, the lighting considered suitable for a factory tool room will place much more emphasis on lighting the task than on the appearance of the room, but in a hotel lounge the priorities will be reversed. This variation in emphasis should not be taken to imply that either task performance or visual appearance can be completely neglected. In almost all situations the designer should give consideration to both these aspects of lighting.

Lighting affects safety, task performance and the visual environment by changing the extent to which, and the manner in which, different elements of the interior are revealed. Safety is ensured by making any hazards visible. Task performance is facilitated by making the relevant details of the task easy to see. Different visual environments can be created by changing the relative emphasis given to the various objects and surfaces in an interior. Different aspects of lighting influence the appearance of the elements in an interior in different ways.

This part of the *Code* discusses the influence of each important aspect of lighting separately. However, it should always be remembered that lighting design involves integrating the various

Figure 1.1 Ensuring the safety of people in the interior

Figure 1.2 Facilitating the performance of visual tasks

Figure 1.3 Aiding the creation of an appropriate visual environment

aspects of lighting into a unity appropriate to the design objectives. This process is discussed in Part 3, Lighting design.

1.2 Daylight and electric light

People prefer a room with daylight to one that is windowless, unless the function of the room makes this impracticable. Few buildings are in fact windowless, but it is also true that in the majority of present-day buildings some of the electric lighting is in continuous use during daytime hours. Electric lighting and daylighting should always be complementary.

The use of daylight with good electric lighting controls can lead to a significant saving in the primary energy used by a building, to national advantage and to the benefit of the environment and building users (Figure 1.4).

A window or rooflight may serve one or more of three main visual purposes: to provide a view, to increase the general bright-

Figure 1.4 For efficient use of energy and for lighting of high quality, the electric lighting and the daylighting should be complementary

ness of a room, and to provide illumination for task performance. These three functions must be considered separately by the designer. A window or electrical installation that serves one purpose well may not be adequate for another – for instance, an opening that provides a good view might give good task lighting but not enhance the general appearance of the room.

Recommendations for daylighting and supplementary electric lighting are given in *BS 8206* Part 2.

1.2.1 Providing a view

A room that does not have a view to the outside, and where one could reasonably be expected, will be considered unsatisfactory by its users. Unless an activity requires the exclusion of daylight, a view should be provided. Sometimes a view is essential for security or supervision, but all occupants of a building should have the opportunity of the refreshment and relaxation offered by a change of scene and focus. Even a limited view to the outside is valuable. If this is not possible, an internal view possessing some of the qualities of an outdoor view could be made available – into an atrium, for example. Sometimes a view into a room is required, for display or for security. More often there is a need for privacy, and this must be taken into account when windows are planned for an external view.

The design of windows for view is covered in the *Lighting Guide 10: Daylighting and Window Design.*

1.2.2 Increasing general room brightness

A user's perception of the character of a room is related to the brightness and colour of all the visible surfaces, inside and outside. The general lighting in a room is a separate consideration from the task illumination, but is equally important. It can be achieved by using daylight or electric light, or both, but the natural variation of daylight is valuable. The light from a side window, in particular, enhances the architectural modelling of a room, and its variation with time gives information about the weather and the time of day.

The character of a naturally lit room is often considered valuable by users. A room can appear daylit even though the principal illumination on the working plane is from electric sources. Contrast between inside and outside is reduced when there is a high level of diffuse daylight internally and when light from luminaires falls on the walls and ceiling. The detailed design of the window frames or surrounds is also important.

Provided that it does not cause thermal or visual discomfort, or deterioration of materials, direct sunlight is appreciated by users. It is especially welcomed in habitable rooms used for long periods during the day, and in buildings where occupants have little direct contact with the outside (such as those used by the elderly). Good control of the sunlight is, however, essential, particularly in working interiors. Generally, sunlight should not fall on visual tasks or directly on people at work. Criteria for window sizes to achieve good general lighting are given in section 2.2, Recommendations for daylighting. Direct sunlight is covered in detail in *Lighting Guide 10: Daylighting and Window Design*.

1.2.3 Illumination for task performance

When there are visual tasks to be carried out, the principles of lighting design using daylight are the same as those for electric lighting: it is necessary both to achieve a given quantity of illumination and to take account of the circumstances that determine its quality. Daylight has particular characteristics as a task illuminant:

(*a*) A constant illuminance on the task cannot be maintained. When the sky becomes brighter, the interior illuminance increases and, although control is possible with louvres, blinds and other methods, fluctuations cannot be avoided. Conversely, in poor weather and at the end of the working day, daylighting may need to be supplemented with electric lighting.

(*b*) Windows, acting as large diffuse light sources to the side of a worker, give excellent three-dimensional modelling. Rooflights, which give a greater downward lighting component, give similar modelling to large ceiling-mounted luminaires.

(*c*) The spectral distribution of daylight varies significantly during the course of a day, but the colour rendering is usually considered to be excellent.

(*d*) When tasks are seen in the same field of view as the bright sky, performance can be impaired by disability glare. If surfaces are placed so that the view of the window is mirrored in them (as when pictures are on a wall which faces a window), visibility can be impaired by the glossy reflections.

The use of windows to provide task lighting in working interiors is economically valuable in many buildings, but the success is dependent on good control of the electric lighting. This is described in section 3.7, Energy management, and in Lighting controls (see CD).

1.3 Lighting levels

The human eye can only perceive surfaces, objects and people through light that is emitted from them. Surface characteristics, reflection factors, and the quantity and quality of light determine the appearance of the environment.

These variables create unlimited permutations between the physical elements and the light that strikes them. Nevertheless, when dealing with an interior it is useful to quantify the luminous flux received per unit of area – i.e. the illuminance measured in lumens per square metre, or lux. The illuminance can be specified and measured as planar, scalar, cylindrical and vector illuminance. These are explained elsewhere in this *Code* (Alternative calculations of illuminance, and Verification of lighting installation performance – see CD). The commonly used planar illuminance relates to tasks that lie in a horizontal, inclined or vertical plane. The plane within which the task is seen is called the reference plane. It is assumed that many critical tasks take place on the flat surface of a desk or bench, and this establishes a horizontal reference plane at the height of the desk or bench tops. This is referred to as the working plane.

This *Code* deals principally with recommendations relating to the task(s), and requires that each task is correctly illuminated and that extreme variation is avoided both across the task and within the space. The illuminance of the immediate surrounding areas should be related to the illuminance of the task area, and should provide a well-balanced luminance distribution in the field of view. For the sake of convenience the recommendations are often applied to the entire working plane, but the designer should be aware of the many tasks that do not lie on the horizontal plane and therefore require separate consideration (see section 1.3.2, Satisfaction).

Measures of illuminance are important because they influence three key aspects of the visual environment: task performance, satisfaction and appearance.

1.3.1 Task performance

The ability to see degrees of detail is substantially determined by size, contrast and the viewer's vision. Improvement to lighting quantity and quality makes an important contribution to improved visual performance. The effect of lighting on task performance is illustrated in Figures 1.5 and 1.6.

Three important points should be noted:

- increasing the illuminance on the task produces an increase in performance following a law of diminishing returns
- the illuminance at which performance levels off is dependent on the visual difficulty of the task – i.e. the smaller the size and the less the contrast of the task, the higher the illuminance at which performance saturates
- although increasing illuminance can increase task performance, it is not possible to bring a difficult visual task to the same level of performance as an easy visual task simply by increasing the illuminance.

Figure 1.5 The effect of lighting on task performance depends on the size of the critical details of the task and on the contrast with their background

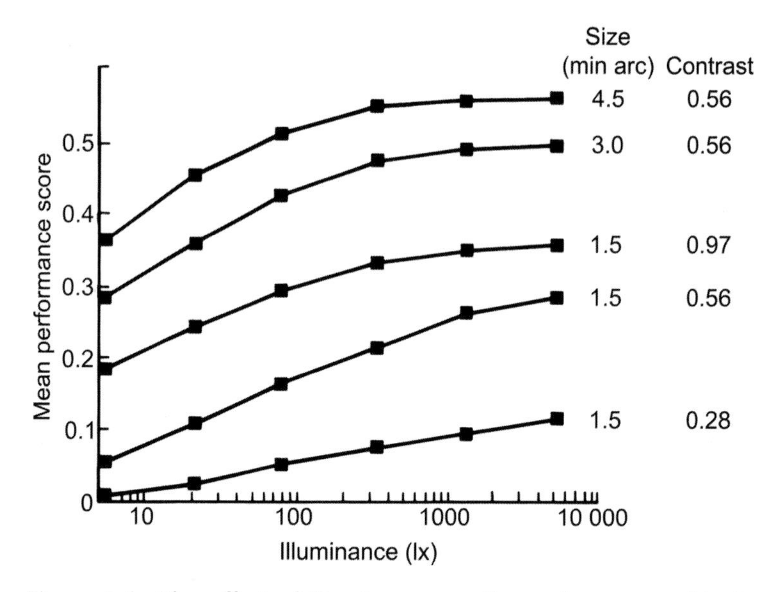

Figure 1.6 The effect of illuminance on the performance of tasks carried out under laboratory conditions

In principle these effects occur for all tasks, although the exact relationship between the illuminance on the task and the performance achieved will vary with the nature of the task. Another aspect is the extent to which the visual part of the task determines the overall performance. Where there is only a small visual component, as in audio typing, the influence of illuminance on overall task performance is likely to be small; however, where the visual component is a major element of the complete task, as in copy typing, the illuminance provided will have a greater influence.

1.3.2 Satisfaction

Subjective response to a space depends on more than task illuminance, and the adjectives that express such responses include 'bright', 'dull', 'gloomy', 'under-lit' and 'well-lit'. The spatial distribution of light, particularly on vertical surfaces, determines these reactions and influences adaptation (see section 1.4.6, Adaptation), which affects visual performance. The ratios between task, wall and ceiling luminances have a strong influence on satisfaction (see sections 2.3.4, Luminance and illuminance ratios, and 2.3.5, Room surfaces).

Figure 1.7 shows mean assessments of the quality of lighting obtained in an office lit uniformly by a regular array of luminaires. Increasing the illuminance on the plane of the desk increases the perceived quality of the lighting, until it saturates at about 800 lux. This demonstrates the importance of the illuminance as one factor in determining people's satisfaction with an interior.

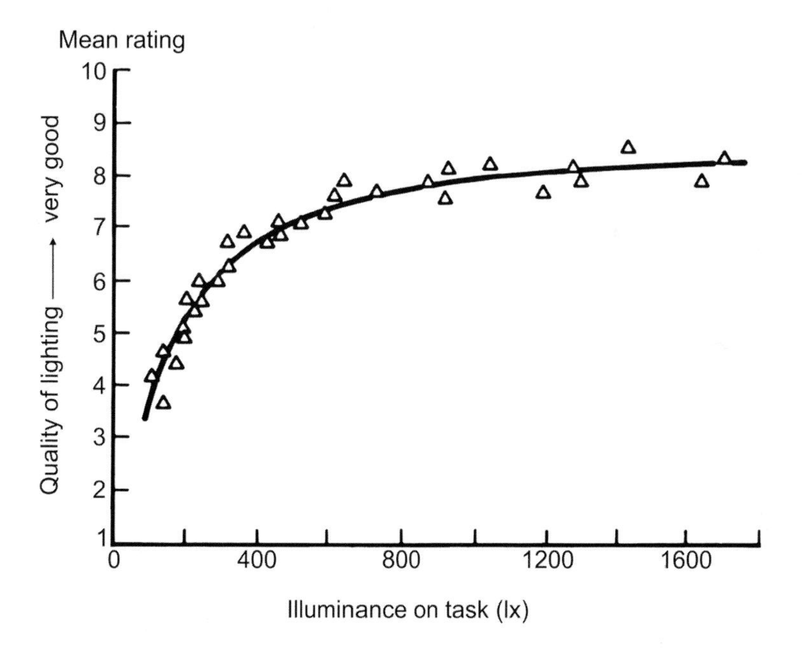

Figure 1.7 Mean assessments of the quality of lighting obtained in an office lit uniformly by a regular array of luminaires

There is no sharp cut-off where lighting conditions move from 'bad' to 'good'. Figure 1.7 shows that as illuminance increases from a low level there is initially a rapid improvement, but as illuminance continues to increase the improvement becomes smaller, until eventually it ceases altogether. So, identifying a suitable illuminance for an interior is a matter of judgement.

The recommended 'standard maintained illuminance' on an appropriate plane for each specific application is given in section 2.5, Lighting schedule. This is converted to the 'design maintained illuminance' by referring to section 2.3.2, Illuminance.

It should be noted that not all working planes are horizontal. Figure 1.8(a) shows vertical task lighting in an art gallery, while Figure 1.8(b) shows vertical task lighting in a supermarket.

1.3.3 Appearance

Any space can be revealed in a variety of ways, and the degree of visual stimulus will depend on the use(s) of the space. Some lighting, especially in non-working environments, will not have a direct, task-related function. Such lighting will express the

Figure 1.8 (a) Vertical task lighting in an art gallery; (b) vertical task lighting in a supermarket

architecture, create appropriate mood, provide emphasis and establish visual coherence. Integrating these non-functional lighting elements within the total lighting design and deciding how to interpret the architecture requires the designer to move beyond pure engineering considerations, taking account of form, colour, texture and architectural intent. To light a space in a manner that is sympathetic to changes in daylight, function and mood will require the designer to anticipate such changes and develop an appropriate lighting solution.

1.4 Variation in lighting

1.4.1 Illuminance variation: definition

When applied to lighting, 'variation' can be in either time or space and can have at least three meanings:

(*a*) Short-term variation occurs either naturally with daylight or with controllable lighting equipment that may change automatically, prompted by changes in daylighting in response to various signals or user manual control.

(*b*) Long-term variation occurs as a result of light loss as lamps age and dirt accumulates over a period of months. Some modern lighting control equipment can counteract this effect.

(*c*) Spatial variation means the uniformity or diversity of illuminance over the task and room surfaces throughout an interior space. This can also include the gradation of light revealing texture or the form of objects.

Topics related to (a) and (b) are discussed in Part 3, Lighting design, and also in Lighting equipment (see CD). Spatial variation will be discussed in more detail in the following sections, with the terms 'uniformity' related to variation in illuminance over the task area and 'diversity' to changes throughout the interior.

Note: For a more formal definition of some of the terms, involved see Part 4, Glossary.

1.4.2 Spatial variation of illuminance in working locations

Variation of illuminance can be considered in two areas: on and around the visual task itself, and over the whole interior. The task area may be considered as the area containing those details and objects necessary for the performance of the given activity, and includes the immediate surround (or background) to the details or objects. Excessive rates of change of illuminance over the task can be distracting and cause changes in visual adaptation across the task, which will reduce visual performance. Excessive variations of illuminance within an interior may affect comfort levels and visual performance by causing transient adaptation problems. These problems are partly addressed by other recommendations, such as those governing the wall-to-task and ceiling-to-task illuminance ratios, and the surface reflectance recommendations (see sections 2.3.4, Luminance and illuminance ratios, and 2.3.5, Room surfaces). Excessive variation in horizontal illuminance will also contribute to these problems, and should be avoided (see section 2.3.3, Illumination variation).

General lighting installations lit by ceiling-mounted arrays of luminaires will usually provide acceptable uniformity conditions over the task areas if luminaires are installed within their recommended spacing-to-height ratios as published by lighting manufacturers (see sections 3.6.2, Selection of luminaire characteristics, and 3.8.3.4, Maximum spacing-to-height ratio (SHR_{max})).

Further information on the effects of obstructions and illuminance variation when using local or localised lighting will be found in sections 2.3.3, Illuminance variation, and 3.8.4, Specification and interpretation of illuminance variation, and also in Verification of lighting installation performance (see CD).

1.4.3 Illuminance variation in non-task locations

There are many lighting applications that do not demand the performance of an exacting visual task for long periods. In public and private areas, the lighting design may be required to entertain and stimulate those using the space. In other areas, leisure, relaxation or even contemplation may be required. To achieve this the lighting designer may be justified in introducing more or less variation. This approach is discussed further in sections 1.4.6, Adaptation, and 2.3.8, Modelling and emphasis. However, it is important to remember to ensure that there is sufficient illumination to ensure the safe circulation of people within the space.

1.4.4 Illuminance, luminance and brightness

The calculation and measurement of the amount of luminous flux (lumens) per unit area reaching various surfaces is the basis of most lighting design. This is primarily because illuminance is relatively simple to calculate and measure. The disadvantage is that the visual system responds physiologically to the luminance distribution in the field of view, but does not perceive the image in this way. The viewer is able to interpret the scene by differentiating between surface colour, surface reflectance, and illumination. This process involves the phenomena known as brightness and colour constancy. For example, if a brown wall is illuminated from one side, resulting in a strong gradation of luminance across the wall, it will still be perceived as a wall of constant colour and reflectance with a variation in illuminance across it. If constancy did not apply, the colour of the wall would appear to change.

The quantity, luminance, depends on both the illuminance and the reflectance of the surface. Illuminance and luminance are both objective quantities but neither relate directly to the subjective response to the 'brightness', which is what the eye and brain 'see' (Figure 1.9).

Luminance, however, provides an important objective link between the illuminance provided and the apparent brightness of the scene.

Note: For a more formal definition of some of the terms involved, see Part 4, Glossary.

1.4.5 Luminance in the visual field

The lighting system will produce patterns of luminance over the task, the immediate surroundings and the peripheral field of view. For reasons of visual satisfaction, comfort and performance, the luminances within the visual field need to be correctly balanced. Too low a luminance surrounding a critical visual task that involves high-reflectance white paper can produce uncomfortable viewing conditions. Low-reflectance office desk tops can certainly give rise to this problem (see section 2.3.5.3, Floor and working plane). Similarly poor viewing conditions can result from the reverse situation of too high a luminance alongside the visual task, when for example the use of a high-reflectance wall finish in an art gallery reduces the detail that can be seen in a dark-coloured painting.

Figure 1.9 (a) Luminance is an objective quantity; (b) brightness is a subjective experience

The lumen method (see section 3.8.3, Average illuminance (lumen method)) of design provides the average horizontal plane illuminance at the floor or working plane, and can be extended to give average illuminance values over the walls and ceiling. Examples of illuminance ratios between the task and the walls or ceiling are given in section 2.3.4, Luminance and illuminance ratios, for typical office workplace lighting. However, when lighting the architectural structure is the main design objective, these illuminance ratios need not apply. The average illuminance can be converted to average luminance values by applying the mean reflectance of the main room surfaces. This gives no detail of the luminance pattern in the field of view. Point by point computation methods (Calculations guide, see CD) with data on the reflectance characteristics of all relevant surfaces and objects can be used to predict the more detailed and complex luminance pattern. To interpret these results in terms of the visual appearance produced, account must be taken of the visual mechanism known as adaptation.

1.4.6 Adaptation

The subjective visual appearance will depend upon adaptation, which is governed by the luminances of the various elements within the field of view, the sizes of the areas involved, and their location with respect to the lines of sight of observers. Levels of adaptation continually change as the eyes move.

The eye can adapt to a wide range of lighting conditions. For example, headlines in a newspaper can be read under moonlight (which provides some 0.2 lux), or by daylight (where the illuminance may be of the order of 100 000 lux). However, the eye cannot adapt to the whole of this range at one time. At night the headlights of an oncoming car will dazzle a dark-adapted viewer, whereas on a sunny day these lights will be barely noticeable. Inside a room daylit by large windows, conditions might allow all objects and surfaces to be viewed comfortably; however, looking into the room from the outside (when adapted to the bright daylight conditions) the windows will appear black and no internal objects or surfaces will be visible.

Our eyes are drawn to the brightest part of a scene. Within work areas, therefore, higher luminance values usually occur at the task areas, but if this is taken to extremes, brightness constancy may break down. This can be avoided by providing adequate illuminance with good colour rendering and glare control. Sharp shadows, sudden large changes in luminance, and excessively bright and frequent highlights should be avoided.

With a uniform electric lighting system and medium to high reflectances of the main surfaces of an interior, the range of luminance will usually be satisfactory. Light ceilings and floors will ensure a high proportion of inter-reflected light and will avoid dark corners.

Reflectance of room surfaces strongly affects the perceived atmosphere in the room. In section 2.3.5, Room surfaces, typical ranges of reflectance are given for major room surfaces.

In comparison with electric lighting, the luminance range produced by sunlight and daylight in an interior will vary enormously. The ranges of luminance in an interior will remain relatively constant for overcast daylight conditions, despite changes in the external illumination, but this will not be the case with sunlight penetration. Here control will be needed, particularly in working areas, because of the adaptation problems that can occur.

Within many working interiors there will be areas intended for circulation or relaxation. It may be desirable here to provide a wider range of luminance values for variety and visual stimulation. Some specular reflections and a limited amount of sparkle would be welcomed from sunlight or display lighting.

1.5 Glare

Glare occurs whenever one part of an interior is much brighter than the general brightness in the interior. The most common sources of excessive brightness are luminaires and windows, seen directly or by reflection. Glare can have two effects: it can impair vision, in which case it is called disability glare (Figure 1.10), and it can cause discomfort, in which case it is called discomfort glare (Figure 1.11). Disability glare and discomfort glare can occur simultaneously or separately.

1.5.1 Disability glare

Disability glare is most likely to occur when there is an area close to the line of sight that has a much higher luminance than the object of regard. Then, scattering of light in the eye and changes in local adaptation can cause a reduction in the contrast of the object. This reduction in contrast may be sufficient to make important details invisible, and hence may influence task performance. Alternatively, if the source of high luminance is viewed directly, noticeable afterimages may be created. The most common sources of disability glare indoors are the sun and sky seen through windows (see previous section) and electric light sources seen directly or by reflection (Figure 1.12). Care should be taken to avoid disability glare in interiors by providing some method of screening windows and avoiding the use of highly specular surfaces.

Figure 1.10 Disability glare from bright sky in front of a VDT makes the screen difficult to read

Figure 1.11 Discomfort glare from bright luminaires

As disability glare is caused by excessive luminance in the field of the view it should be avoided by using luminaires which give suitable shielding of their lamps. Table 1.1 gives a list of minimum shielding angles for a given lamp luminance.

1.5.2 Discomfort glare from electric lighting

The discomfort experienced when some elements of an interior have a much higher luminance than others can be immediate, but sometimes only becomes evident after prolonged exposure. The degree of discomfort experienced will depend on the luminance and size of the glare source, the luminance of the background against which it is seen, and the position of the glare source relative to the line of sight. A high source luminance, large source

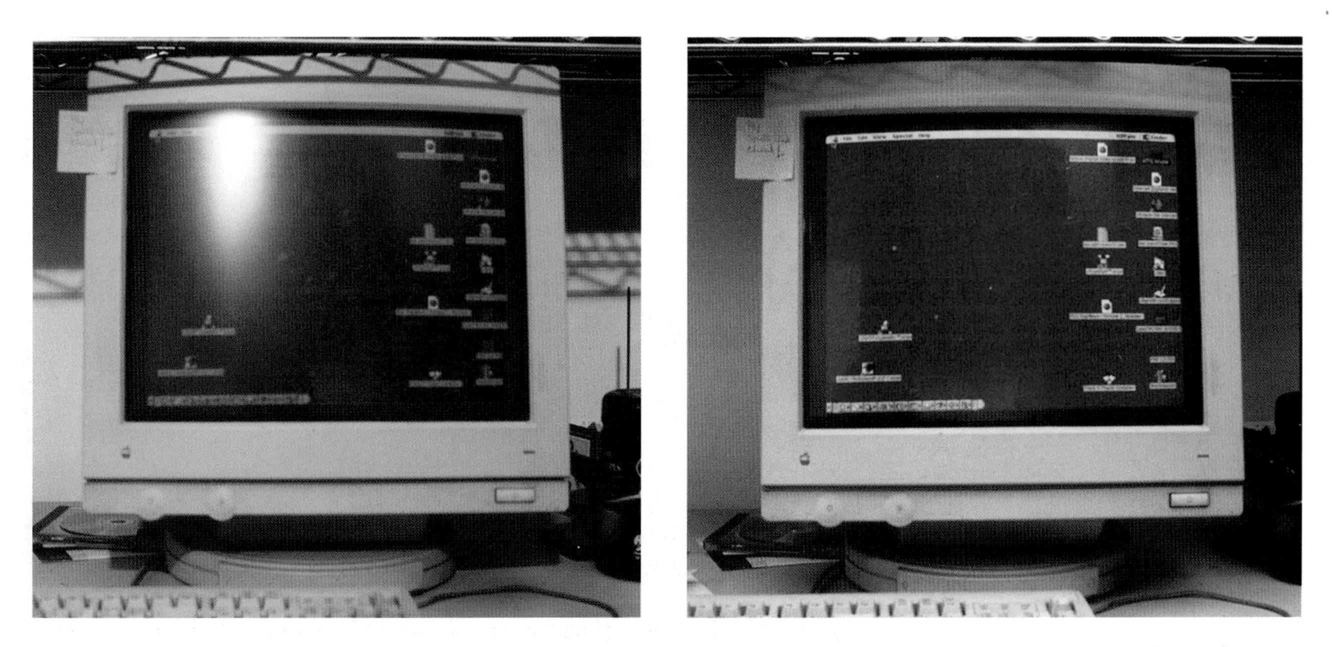

Figure 1.12 (a) Effect of veiling reflections from electric lighting on a VDT screen; (b) VDT screen without veiling reflections

Table 1.1 Lamp luminance and shielding

Lamp luminance (kcd/m^{-2})	Minimum shielding angle (°)
1 to < 20	10
20 to < 50	15
50 to < 500	20
≥ 500	30

area, low background luminance and a position close to the line of sight all increase discomfort glare. Unfortunately, most of the variables available to the designer alter more than one factor – for example, changing the luminaire to reduce the source luminance may also reduce the background luminance. These factors could counteract each other, resulting in no reduction of discomfort glare. However, as a general rule, discomfort glare can be avoided by the choice of luminaire layout and orientation, and the use of high-reflectance surfaces for the ceiling and upper walls. In any proposed lighting installation, the likelihood of discomfort glare being experienced can be estimated by calculating the unified glare rating (UGR) (see section 3.8.5, Discomfort glare; see also Calculation of discomfort glare, see CD). Recommended limiting glare ratings for specific applications are given in section 2.5, Lighting schedule, and for luminous ceilings and indirect lighting installations maximum luminances are given in section 2.3.5.1, Ceilings.

1.5.3 Discomfort glare from windows

Severe visual discomfort arises when a person is looking through a window in the direction of the sun, or when direct sunlight falls on a light-coloured surface in the immediate field of view. In such circumstances there may also be thermal discomfort. Solar control is essential in most buildings; this may be in the form of the design of the building's overall form and orientation, or the use of external screens and louvres, glass of low transmittance, or internal blinds and curtains. All of these reduce the total amount of light entering a room, and this must be considered by the lighting designer.

Glare can also arise when an overcast sky is viewed through a window. It may be reduced by solar control devices, or by other means of decreasing the contrast between the interior and the view of the sky. These include the use of splayed window reveals to give areas of intermediate brightness, ensuring that the window wall is light-coloured, and using electric lighting to increase the luminance of the window wall.

1.5.4 Veiling reflections

Veiling reflections (Figure 1.13) are high-luminance reflections that overlay the detail of the task. Such reflections may be sharp-edged or vague in outline, but regardless of form they may affect task performance and cause discomfort. Task performance will be affected because veiling reflections usually reduce the contrast of a task, making task details difficult to see, and may give rise to discomfort.

Figure 1.13 Veiling reflections in an industrial task

Two conditions have to be met before veiling reflections occur:

- part of the task, task detail or background, or both, has to be glossy to some degree
- part of the interior, called the 'offending zone', which specularly reflects towards the observer has to have a high luminance.

The most common sources of veiling reflections are windows and luminaires. Generally applicable methods of avoiding veiling reflections are the use of matt materials in task areas, arranging the geometry of the viewing situation so that the luminance of the offending zone is low, or reducing the luminance by, for example, using curtains or blinds on windows.

It should be noted that although veiling reflections are usually detrimental to task performance, there are some circumstances in which they are useful. *Lighting Guide 1: The Industrial Environment* contains examples of the use of high-luminance reflections in inspection lighting (see CD).

1.6 Directional qualities and modelling

The direction and distribution of light within a space substantially influence the perception of the space as well as objects or persons within it. Decisions that determine such perception relate partly to the provision of desirable illuminance values and partly to the subjective issues of architectural interpretation, style and visual emphasis. Good lighting design results from an appreciation both of the nature and qualities of the surfaces upon which light falls, and of the methods of providing such light. The visual characteristics of surfaces and sources of light are interrelated and interdependent. The appearance of a surface or object will depend on the following:

(*a*) Its colour and reflectance, and whether it is specular or diffuse, smooth or textured, flat or curved. All surfaces reflect some portion of the light falling on them and so become sources of light. Depending on their degree of specularity, texture and shape, their appearance will also vary with the direction of view.

(*b*) The layout and orientation of luminaires and sources of reflected light. Single sources of relatively small size will produce harsh modelling, the effect becoming softer as the number and size of the sources increase. The predominant direction of light has a fundamental effect on appearance; lighting from above provides a distinct character that is totally different to that achieved by lighting from the side or from lower angles. In addition, colour differences between sources of light of various distributions and orientation strongly influence the lit appearance of spaces, surfaces and objects. With so many variables, luminance patterns become too complex to predict in detail.

This element of unpredictability is generally acceptable (or even desirable) provided that the basic rules of good lighting practice are observed, such as the limitation of extremes of glare, contrast and veiling reflection. The importance of modelling is obvious for retail display, exhibition work and the creation of mood. However, any lighting installation that fails to create appropriate degrees of modelling will provide visual results that are perceived as bland and monotonous. Virtually all environments can benefit from a lighting approach that considers the question of direction and the resulting revelation of architectural form, texture and facial modelling. The designer must decide where in the range, from harsh or dramatic to soft or subtle, modelling the design aim should be set. Further information is given on specification of modelling in section 2.3.8, Modelling and emphasis, and on modelling design in section 3.6.3, Illuminance ratio charts.

1.6.1 Revealing form

The revelation of the form of an object or structure is determined by the relationship of the incident angle and intensity of light to the surface in question, the position of the viewer relative to the surface, and the nature or composition of the surface.

Light reveals surfaces by three basic methods; emission, silhouette and reflection. Figure 1.14(a)–(e) show an identical form revealed by these methods. Revelation by emission (Figure 1.14(a)) or silhouette (Figure 1.14(b)) exposes little or none of the three-dimensional quality of the form. However, the white vase (Figure 1.14(c)) is dramatically revealed as three-dimensional by the gradation of reflected light over its surface. The same visual message is provided by the highlight on the surface of the glossy black vase (Figure 1.14(d)). The vase in Figure 1.14(e) is lit to provide a balanced rendering of the form by the use of a strong rear 'key' light and a less intense frontal 'fill' light.

The relationship between the intensity of strongly directional, emphatic lighting and the ambient or general illuminance level within a space is critical. In an otherwise dark space a relatively low intensity of directional light will strongly reveal an object, whilst the same degree of emphasis in a brightly lit space will require considerably greater intensity from the directional highlighting. Subtle and pleasant modelling is usually favoured in general working areas and public spaces, where more extreme

(a)

(b)

(c)

(d)

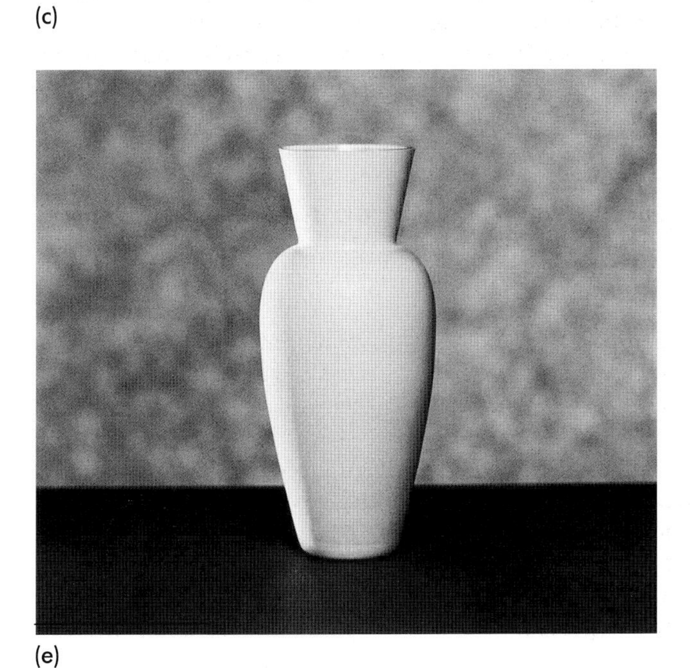

(e)

Figure 1.14 The influence of surface finish and lighting set-up on the appearance of identically shaped vases: (a) frosted glass vase lit internally from below; (b) matt black vase silhouetted against a lighted background; (c) matt white vase lit from the front right at about 45° with back-lighting added from left; (d) glossy black vase, lighting positions the same as used with (c) but using narrow beam spotlights; (e) matt white vase lit from the same positions as (c) but the intensity of the back light has been doubled in intensity, diffused front lighting has been used

ratios, especially when combined with unusual angles of directional light, will produce an increasingly dramatic and distorted effect on faces. However, this may prove appropriate for other circumstances, such as architectural detailing, sculpture, museum artefacts and some types of retail display (see section 2.3.8, Modelling and emphasis).

Since much electric lighting emanates from ceiling locations, it is important to consider the relationship between predominantly vertical downlighting and light reflected from the surrounding walls and floor. Insufficient reflected light will result in harsh facial shadowing. The lighting designer should consider the reflection factor of the walls and their illuminance to ensure a satisfactory balance.

1.6.2 Revealing texture

The revelation of the texture of a material can have both aesthetic and functional value. Figure 1.15 shows an example of lighting used to reveal surface texture. By lighting at oblique angles, the texture of the shuttered concrete is revealed as an architectural feature of the building. The deliberate use of a non-uniform luminance pattern provides greater visual impact without losing structural coherence. If the spotlights had been directed at near right angles to the surfaces, or diffused lighting had been used, the interior would have lost vitality and interest.

The texture can be suppressed or expressed by applying light at an appropriate angle (Figure 1.15). The decision to reveal the texture or not is one related to style and architectural expression. The functional revelation of texture is illustrated in Figure 1.16, which shows the effect of light at glancing angles over fabric in order to detect a pulled thread.

Figure 1.15 Revealing architectural texture

Figure 1.16 Revealing a pulled thread

Not all revelation of texture is desirable. A common problem arises with uplighting on badly finished ceiling surfaces, which reveals unwanted 'texture' or a degree of unevenness that other angles of light would not reveal. Shadows and highlights can reveal too much textural detail, which can result in a reduction of task visibility; the degree to which texture is revealed should therefore be related to the particular requirements of the task.

1.6.3 Display lighting

The preceding comments about the directional qualities of lighting are particularly relevant to display lighting techniques. Revelation of form, dimension and texture are objectives that are invariably encountered in retail and other forms of display work. Additionally, the question of colour appearance and colour rendering is critical. Figure 1.17(a)–(f) illustrate some of the basic techniques for revealing a three-dimensional model to best effect when viewed from one angle. The illustration shows six basic approaches and the associated optimum incident lighting angles. In practice, numerous combinations of these can be used to achieve the required balance of emphasis and revelation (see section 2.3.8, Modelling and emphasis).

1.7 Surfaces

The effect a lighting installation creates in an interior is strongly influenced by the properties of the major room surfaces. For this reason, if for no other, the lighting designer should always attempt to identify the proposed surface finishes early in the design process. The main properties of the room surfaces that are relevant to the appearance of the space are their reflectance and their colour.

1.7.1 Surface reflectances

For interiors lit from the ceiling, the significance of the ceiling reflectance increases as the room area increases. In a small room,

(a) Low-level side lighting from the right

(b) High-level side lighting from the left at approximately 45°

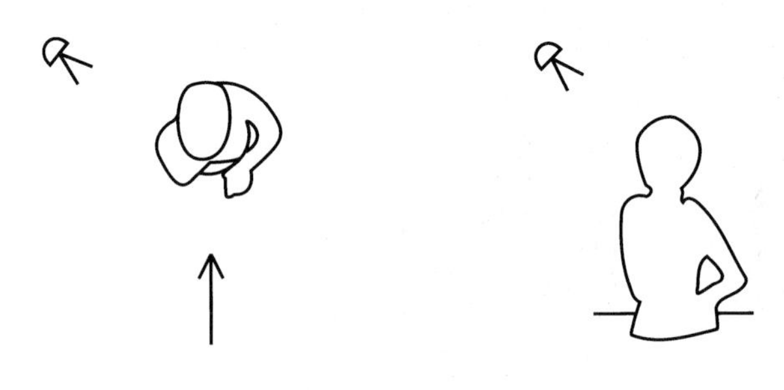

(c) Combination of high intensity 'key' lighting from the front left of the picture at about 40° and a diffused 'fill' light of lower intensity from the right

Figure 1.17 Examples of the modelling effects that can be produced by some basic display lighting techniques. These illustrations are limited to one viewing direction and the lighting of the background is unchanged. The diagrams to the right of each picture indicate the lighting arrangement used

(d) 'Backlighting' from above and slightly to the right

(e) Front lighting at a low angle from the right

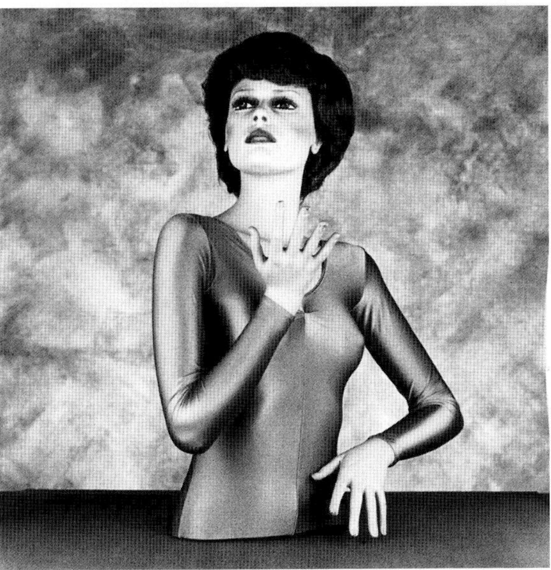

(f) Diffused lighting from the front and sides

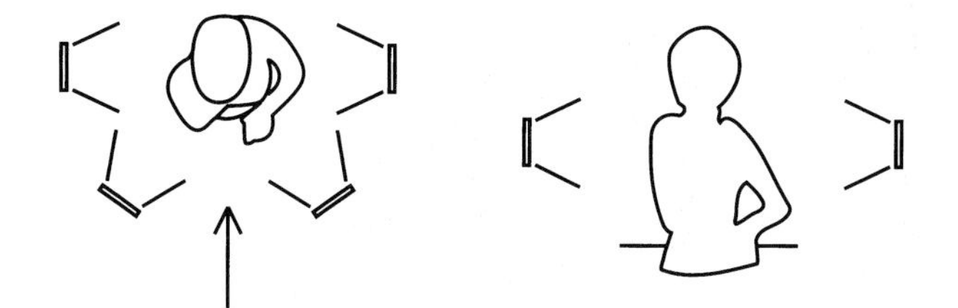

the ceiling is not conspicuous and its contribution to the illuminance on the working plane is usually small. In a big room, the contribution of light reflected from the ceiling to the total illuminance on the working plane is usually large and the ceiling occupies a substantial proportion of the visual field. Achieving an acceptable reflectance for the ceiling cavity requires a white or near-white ceiling. In small rooms a low-reflectance ceiling may be acceptable, although if the room is predominantly lit by daylight from side windows the room may appear gloomy if too low a reflectance is chosen. Where indirect lighting is used, a white or near-white ceiling is essential, regardless of room size.

Wall reflectance is usually unimportant to the lighting of a large room except for positions close to the wall. If low wall reflectances are used, the illuminance in the adjacent areas may be too low. In small rooms, wall reflectance is always important. High wall reflectances will enhance the illuminance on the working plane and increase the inter-reflected component of the lighting, thereby improving uniformity. The importance of having a high wall reflectance is increased when the room is predominantly lit by daylight from side windows. In all rooms, unless a high-reflectance finish is applied to the window wall, the luminance difference between the window wall and the daytime view through the window may be excessive and uncomfortable.

All this suggests that a high-reflectance finish to walls is highly desirable. However, the use of high-reflectance wall finishes should be treated with caution. Large areas of high reflectance may compete for attention with the task areas, leading to eyestrain and feelings of discomfort. Furthermore, if the high-reflectance surfaces are produced using gloss paint, reflected glare is likely to occur. The effective reflectance of the wall finish will be reduced by windows, unless light-coloured blinds or curtains are used. Dark wall hangings, cupboards or other equipment above the working plane will also reduce the effective wall reflectance. Where the perception of people's faces is important, for example in lecture theatres and conference rooms, the brightness of the walls needs to be controlled as these form the background against which people are seen.

Dark floor cavities will tend to make ceilings and walls look underlit, especially when daylight from side windows is used; however, using very light floors tends to create a maintenance problem. Recommendations for room surface reflectance are given in section 2.3.5, Room surfaces, and the effect on installed load is discussed in section 2.4, Energy efficiency recommendations.

1.7.2 Surface colours

Surface colour can be classified by the use of a colour system, which allows colour to be specified unambiguously. For the purposes of lighting design and calculation, information on the reflectance of surface colours is required. Several colour systems exist, some of which can be used to estimate reflectance. Further information on the most commonly used systems is given in *Lighting Guide 11: Surface Reflectance and Colour*. In the Munsell system, for example, each colour is specified by three quantities: hue (whether a colour is basically red, yellow, green, blue, purple etc.), value (the lightness of the colour, related to its reflectance), and chroma (the strength of the colour). This classification forms a convenient basis on which to discuss the effects of room surface colour on the appearance of space.

By choosing different values for different components of the interior it is possible to dramatise or to buffer the pattern of light and shade created by the lighting. An example of this is the use of a high-reflectance (high value) wall opposite a window wall.

By choosing colours of different chromas it is possible to create a pattern of emphasis. Strong emphasis requires strong chromas, but their use calls for caution. An area of awkward shape that might pass unnoticed at weak chroma can look unsightly at strong chroma. Also, a small area of strong chroma might be stimulating but the same chroma over a large area could be overpowering.

The selection of hue is partly a matter of fashion and partly a matter of emotion. By choosing a predominant hue for a space it is possible to create a 'cool' or 'warm', 'restful' or 'active' atmosphere. Figure 1.18 illustrates the use of surface colours in public and commercial interiors. The children's room in the library in Figure 1.18(a) uses upholstery in strong primary colours to provide a vibrant and stimulating atmosphere. This contrasts with the use of blues and reds in the commercial interior in Figure 1.18(b), where a calmer and more sophisticated ambience is required.

All rooms will have a mixture of colours, and this fact raises the question of colour harmony. There are a number of so-called rules of colour harmony, which are little understood. However, it is widely believed that the main variable influencing pleasant colour harmonies is the difference in value for the two colours compared; the greater the difference in value, the greater the chances of achieving a pleasant colour combination. The effect of chroma differences is thought to be similar, combinations of colours with large differences in chroma tending to be pleasant. As for hue differences, there is not believed to be any consistent effect, with all the same hues, closely related hues or complementary

Figure 1.18 (a) Use of colour to enhance the appearance of a children's play area; (b) use of colour to enhance the appearance of a commercial interior (advertising agency meeting room)

hues being capable of creating either pleasant or unpleasant colour combinations.

These observations suggest that when selecting colours for an interior the first aspect to consider is the value of colours, then the chroma and finally the hue. However, once the pattern of light, shade and emphasis has been established by the choice of the value and chroma for different surfaces, the range of hues that is available may be limited. For example, if a given surface is to have both strong chroma and high value, then it must inevitably have a yellowish hue. Conversely, when a surface is required to have low value and strong chroma, a colour from the red to blue part of the hue circle must be used. Once the level of chroma is reduced from a high level, the whole range of colours is available.

The light reflected from a surface of strong chroma will be coloured, and may influence the colour of other surfaces. The most common situation where this is seen is the case of a floor covering of strong chroma lit by a lighting installation that does not light the ceiling directly. In this situation the ceiling will mainly be lit by light reflected from the floor, which will tend to colour the ceiling.

1.7.3 Object colours

The colours of objects within an interior can have a marked effect on the appearance of the space. In choosing a combination of colours for both the surfaces and equipment within a space, it is preferable if the elements can be considered as a whole so that a degree of visual co-ordination can be achieved. The actual choice of a combination of colours to produce a co-ordinated colour scheme is probably one of the most elusive design tasks, and at present there is no single widely accepted design procedure.

There are limitations to the choice of colours of some objects within the space. These arise from the use of colour for the coding of services and to indicate potential hazards. The use of colour for the coding of services is governed by *BS 1710* and should be undertaken sparingly, with emphasis given to identification of outlets, junctions and valves. The use of colour to identify potential hazards is governed by *BS 5378*. Care should be taken to avoid confusion between *BS 5378* on hazard warning colours, *BS 1710* on service colours, and other colours in the interior. The lighting should not unduly distort the colours reserved for services or hazard indication in such a way as to be confusing.

1.8 Light source radiation

This *Code* is primarily concerned with light source radiation in that small part of the electromagnetic spectrum, from 400 nm to 780 nm, which stimulates the sense of sight and colour. However, all light sources radiate energy at shorter wavelengths in the ultraviolet as well as at longer wavelengths in the infrared parts of the spectrum. This radiation can promote physiological effects that are either a benefit or a hazard. The basic function of luminaires is to control the visible radiation (light), but they can also concentrate, diffuse or attenuate the non-visible radiation from lamps. The lighting designer needs to be aware of the effects of all the radiation that is being emitted.

Light has two colour properties; the apparent colour of the light that the source emits, and the effect that the light has on the colours of surfaces. The latter effect is called colour rendering.

1.8.1 Apparent colour of emitted light

The colour of the light emitted by a near-white source can be indicated by its correlated colour temperature. Each lamp type has a specific correlated colour temperature, but for practical use the correlated colour temperatures have been grouped into three classes by the Commission Internationale de l'Eclairage (CIE) as shown in Table 1.2.

Table 1.2 Colour appearance and colour temperature

Colour appearance	Correlated colour temperature
Warm	Below 3300 K
Intermediate	3300–5300 K
Cool	Above 5300 K

The choice of an appropriate colour appearance of a light source for a room is largely determined by the function of the room. This may involve such psychological aspects of colour as the impression given of warmth, relaxation, clarity etc., and more mundane considerations such as the need to have a colour appearance compatible with daylight and yet to provide a 'white' colour at night.

Note: For a more formal definition of some of the terms involved see Part 4, Glossary.

1.8.2 Colour rendering

The ability of a light source to render colours of surfaces accurately can be conveniently quantified by the CIE general colour-rendering index. This index is based on the accuracy with which a set of test colours is reproduced by the lamp of interest relative to how they are reproduced by an appropriate standard light source, perfect agreement being given a value of 100. The CIE general colour-rendering index has some limitations, but it is the most widely accepted measure of the colour-rendering properties of light sources.

Lamps with a colour-rendering index below 80 should not be used in interiors where people work or stay for longer periods. Exceptions may apply for some places or activities (e.g. high-bay lighting), but suitable measures should be taken to ensure lighting with higher colour rendering at fixed, continually occupied workplaces, and where safety colours have to be recognised.

For recommendations, see section 2.5, Lighting schedule.

Note: For a more formal definition of some of the terms involved, see Part 4, Glossary.

1.9 Light modulation

All electric lamps operated on an AC supply (50 Hz in Europe) have an inherent modulation in light output at twice the supply frequency (see Figure 1.19(a)). With most discharge lamps there is also a small component at the supply frequency itself, which can increase as the lamp ages (see Figure 1.19(b)). The 100 Hz modulation in light output is not perceptible by the great majority of people, and the light appears steady. Incandescent lamps only show a small modulation because of thermal inertia; discharge lamps can show a modulation between 17 and 100 per cent.

If lamps with a large modulation are used to light rotating machinery, coincidence between the modulation frequency and the frequency of rotation may cause moving parts to appear stationary. This is called the stroboscopic effect, and can be dangerous (see section 3.6.1, Selection of lamp characteristics).

Light modulation at lower frequencies (about 50 Hz or less), which is visible to most people, is called flicker. Flicker is a source of both discomfort and distraction, and may even cause epileptic seizures in some people. Sensitivity to flicker varies widely between individuals. The perceptibility of flicker is influenced by the frequency and amplitude of the modulation and the area

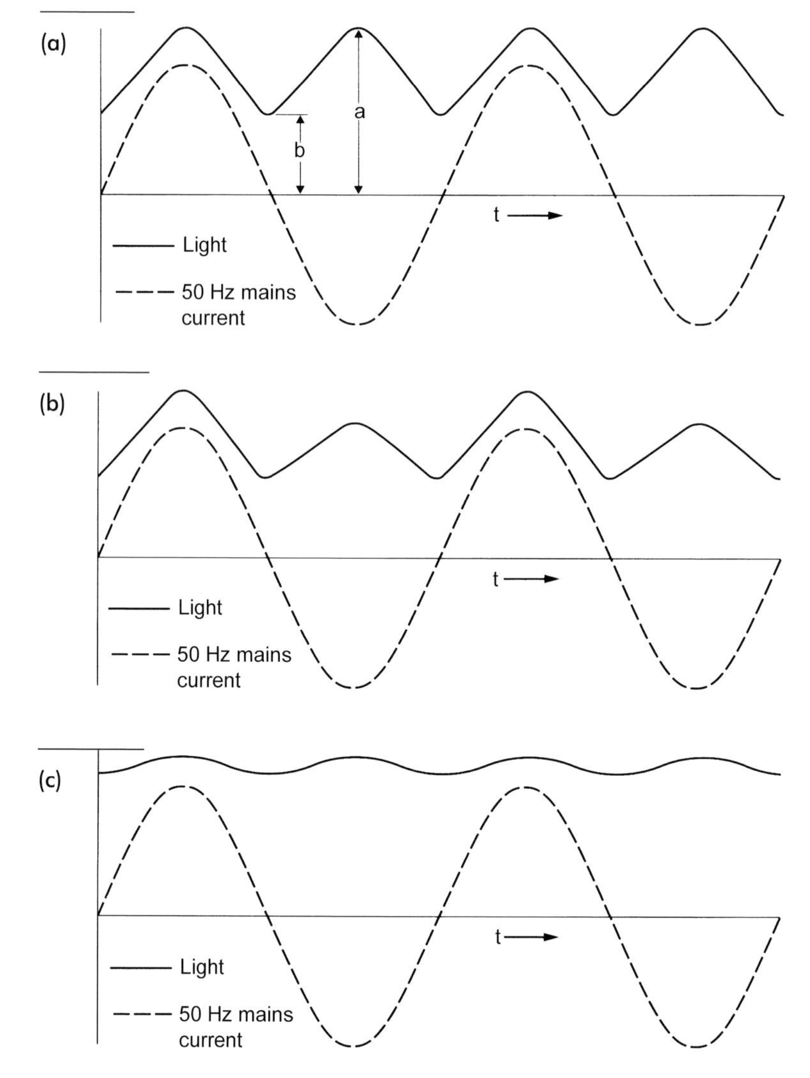

Figure 1.19 Light modulation. (a) 100 Hz light output waveform from typical fluorescent lamp operating on conventional wire wound control gear; (b) changes in electrode characteristics at the end of life can produce 50 Hz ripple on 100 Hz output waveform; (c) 7 per cent modulation of 100 Hz light output of fluorescent lamp operating on high frequency electronic ballast connected to 50 Hz mains supply. The high frequency lamp current is not shown

of vision over which it occurs. For a given set of circumstances, the frequency above which the alternation of visual stimuli is no longer perceptible is known as the fusion frequency. Large amplitude variations over large areas at low frequencies give the most uncomfortable conditions. The eye is most sensitive to flicker at the edge of the field of view; thus visibly flickering overhead lights can be a source of great discomfort.

Lamps driven by high-frequency power supplies (e.g. 35 kHz) overcome these drawbacks in that all significant low frequency modulation below 100 Hz is eliminated. Although 100 Hz modulation of appreciable amplitude does affect a very small minority of the population, the modulation from well-designed electronic ballasts (Figure 1.19(c)) is similar in shape and amplitude to that of incandescent sources.

Note: For a more formal definition of some of the terms involved, see Part 4, Glossary.

Part 2 Recommendations

2.1 Introduction

For good lighting practice it is essential that, in addition to the required illuminance, qualitative and quantitative needs are satisfied.

Lighting requirements are determined by the satisfaction of three basic human needs:

- visual comfort, providing the users of the building with a feeling of well-being – in an indirect way this also contributes to a high productivity level
- visual performance, where the users of the building are able to perform their visual tasks, even under difficult circumstances and for longer periods
- safety.

The main parameters determining the luminous environment are:

- luminance distribution
- illuminance
- glare
- directionality of light
- colour rendering and colour appearance of the light
- flicker
- daylight.

Values for illuminance, discomfort glare and colour rendering are given in section 2.5, Lighting schedule.

The following sections give information on the above topics together with recommendations for energy consumption of lighting.

2.2 Recommendations for daylighting

2.2.1 Daylight for general room lighting

In most types of buildings, users prefer rooms to have a daylit appearance during daytime hours. This appearance can be achieved, even if there is a significant amount of daytime electric lighting, by ensuring that the changing brightness of daylight is clearly noticeable on walls and other interior surfaces. It is also necessary to achieve sufficiently bright interior surfaces to avoid glare from contrast with the sky. In order to control glare from windows, screening should be provided where appropriate.

The following values should be adopted where a daylit appearance is required.

2.2.1.1 Interiors without supplementary electric lighting during daytime

If electric lighting is not normally to be used during daytime hours, the average daylight factor should be not less than 5 per cent.

The internal reflectances and the positions of windows should be such that inter-reflected lighting in the space is strong and even. When the shape of the room causes the distribution of daylight to be very uneven (such as when a large area lies behind the no-sky line – see section 3.1.1, Initial appraisal of daylight quantity), supplementary electric lighting may still be necessary.

2.2.1.2 Interiors with supplementary electric lighting during daytime

If electric lighting is to be used during daytime, the average daylight factor should be not less than 2 per cent.

In a room where the average daylight factor is significantly less than 2 per cent, the general appearance is of an electrically lit interior. Daylight will be noticeable only on room surfaces immediately adjacent to windows, although the windows may still provide adequate views out for occupants in the room.

2.2.2 Daylight for task illumination

Where daylight alone provides the illumination for a visual task, the illuminance should not fall below that given in the Lighting schedule (section 2.5). The uniformity of illuminance within the immediate task area should be similar to that acceptable with electric lighting (see section 2.3.3, Illuminance variation), although there may be differences in the level of daylight in different parts of an interior.

2.3 Recommendations for electric lighting with daylighting

The two distinct functions of electric lighting used in conjunction with daylight are to enhance the general room brightness and to supplement the daylight illuminance on visual tasks (see sections 3.8.1.4, Conventional switching, and 3.8.1.5, Photo-electric control).

Where there is a significant amount of daylight (an average daylight factor of 2 per cent or more), electric lighting may be required to reduce the contrast between internal surfaces and the external view. It needs to fall on the walls and other surroundings of the window opening. The brighter the view, the higher the luminance required of the surfaces surrounding the window. Electric lighting may also be required to increase the general illumination of parts of the room distant from a window. If this is the case, the average working plane illuminance from electric lighting in the poorly daylit areas should not be less than 300 lux. If a lower illuminance is used, in circulation areas for example, there may be noticeable contrast between areas near windows and other parts of the room, with a corresponding impression of harshness or gloominess.

As far as possible, the electric lighting should not mask either the natural variations of daylight across surfaces or the way in which natural lighting changes with time and weather.

When the quantity of daylight in the space is small, the electric lighting is required to give general illumination over all room surfaces. Particular illumination may still be required on surfaces around windows – for instance, in the case of a small pierced window through which an area of bright sky is visible.

Where both daylight and electric light provide task lighting, the combined illuminance should satisfy the criteria given in the Lighting schedule (section 2.5). The directionality of daylight is usually an advantage in achieving good modelling, but electric lighting may be required to increase the luminance of surfaces in shadow. Care should be taken, in the provision of daylight, that tasks are not viewed against the sky or a very bright area of the interior – see Figure 1.10). If this is unavoidable, the background luminance should be such that there is a satisfactory brightness contrast between task and background (see section 2.3.4, Luminance and illuminance ratios).

2.3.1 Colour

The sky varies in colour with time, azimuth and altitude. These variations are very great, and no electric lamp matches continuously the colour appearance of daylight. Whilst there are devices available that can mimic the changing colour of daylight, they are only rarely used to provide artificial lighting. In general room lighting, apparent discrepancies between the colour of electric light and daylight can be reduced by using lamps of intermediate colour temperature (3300–5300 K) and screening them from the view of the occupants, using opaque louvres rather than translucent diffusers.

When discrimination of surface colour is essential for task performance, the choice of lamp should be that recommended for the task under entirely electric lighting (see section 1.8.2, Colour rendering). It may be necessary for the user to know whether the task is illuminated primarily with electric light or with daylight.

2.3.2 Illuminance

The illuminance and its distribution on the task area and its surrounding area have a great impact on how quickly, safely and comfortably a person perceives and carries out a visual task.

The values given in the Lighting schedule (section 2.5) are maintained illuminances over the task area on the reference surface, which may be horizontal, vertical or inclined. The average illuminance for each task shall not fall below the value given in the Lighting schedule, regardless of the age or the condition of the installation. The values are valid for normal visual conditions, and take into account the following factors:

- psycho-physiological aspects such as visual comfort and well-being
- requirements for visual tasks
- visual ergonomics
- practical experience

— safety

— economy.

The value of illuminance may be adjusted by at least one step of illuminance on the scale of illuminances (see below) if the visual conditions differ from the normal assumptions.

A factor of approximately 1.5 represents the smallest significant difference in subjective effect of illuminance. In normal lighting conditions, approximately 20 lx is required to just discern features of the human face, and is the lowest value taken for the scale of illuminances. The recommended scale of illuminances (in lx) is:

20–30–50–75–100–150–200–300–500–750–
1000–1500–2000–3000–5000

The required maintained illuminance should be increased when:

— visual work is critical

— errors are costly to rectify

— accuracy or higher productivity is of great importance

— the visual capacity of the worker is below normal

— the details of the task are of an unusually small size or low contrast

— the task is undertaken for an unusually long time.

The required maintained illuminance may be decreased when:

— task details are of an unusually large size or high contrast

— the task is undertaken for an unusually short time.

It is also assumed that the people doing the work have normal vision. If a significant number of building occupants have some degree of visual impairment, the maintained illuminance could be increased. The most common effects of old age on vision are an increase in the shortest distance at which an object can be focused, reduced light transmission through the eye, and an increase in the scattering of light in the eye. Spectacles or contact lenses can be used to correct the first effect. Increasing the illuminance will offset the loss in transmission and will increase visual sensitivity. A 70-year-old person can require around three times the illuminance needed by a 20-year-old, in order to achieve similar visual performance. The recommendations given in the Lighting schedule (section 2.5) generally assume an age of 40–50 years.

Although increasing the illuminance and avoiding glare will benefit most people with some degree of visual impairment, there are some severe forms of visual defect (e.g. cataract) for which increasing the illuminance may be detrimental. It is essential to match the lighting conditions to the nature of the visual defect.

The Lighting schedule is not intended to cover lighting for the visually handicapped.

2.3.3 Illuminance variation

For the task area and immediate surround, uniformity is important. A task area is not usually the entire area of a workstation. On an office desk, for example, the task area may only be about the size of a desk blotter, but in interiors such as drawing offices the visual task may cover the whole area of a drawing board. The range of task areas is even wider in industry – from a micro-electronics assembly line to a car body production line. However, when the precise size of the task area is not known, calculations can be based on an area measuring 0.5 m × 0.5 m located immediately in front of the observer at the edge of the desk or working surface.

It is recommended that the uniformity of illuminance (minimum to average illuminance) over any task area should not be less than 0.7 (section 3.8.4, Specification and interpretation of illuminance variation; also Measurement of illuminance variation – see CD) and that the average illuminance on the task must be appropriate to that of the activity as set out in the Lighting schedule (section 2.5). Where task areas may be located anywhere over an area of a room, the recommendation applies to all potential task areas within that area. The uniformity recommendation does not necessarily have to apply to the entire room.

The illuminance of the immediate surrounding areas must be related to the illuminance of the task area, and should provide a well-balanced luminance distribution in the field of view. The immediate surrounding area is taken to be a band with a width of at least 0.5 m.

Large spatial variations in illuminance around the task area may lead to visual stress and discomfort.

The illuminance of the immediate surrounding areas may be less than the values in Table 2.1. The uniformity of the surrounding area should be at least 0.5.

Table 2.1 Relationship of illuminances of immediate surrounding areas to task area

Task illuminance (lx)	Illuminance of immediate surrounding areas (lx)
≥ 750	500
500	300
300	200
≤ 200	E_{task}

In most spaces there are various visual tasks with differing degrees of difficulty. Although general lighting systems (see section 3.5.1, General lighting) provide flexibility of task location, the average illuminance is determined by the needs of the most exacting task. It is often wasteful to illuminate all areas to the same level, and non-uniform lighting may be provided by local or localised lighting systems (see sections 3.5.2, Localised lighting, and 3.5.3, Local lighting). If control systems (see section 3.7.1, Choice of controls; also Lighting controls – see CD) are used, individuals may be able to adjust their levels of supplementary task lighting, and presence detection may also switch off luminaires in unoccupied areas. Whatever lighting system is used, excessive variations of horizontal illuminance across an interior must be avoided; the diversity of illuminance expressed as the ratio of the maximum illuminance to the minimum illuminance at any point in the 'core area' of the interior should not exceed 5 : 1. The core area is that area of the working plane having a boundary 0.5 m from the walls (Average illuminance, see CD).

Installations lit by ceiling-mounted arrays of luminaires designed by the lumen method (see section 3.8.3, Average illuminance (lumen method)) following the conventional spacing and layout criteria will usually satisfy the uniformity requirements. It is normal in such installations for the horizontal illuminance at the perimeter to be significantly less than the average value over the working plane. This is particularly marked in interiors having low-reflectance walls (hence the earlier reference to the 'core area'). Local reductions in illuminance, due to shadowing, may also be caused by large items of furniture or equipment that project substantially above the working plane. Both in these areas and at the perimeter of the room, local or localised lighting may be necessary if critical visual tasks are to be performed.

In a localised or local lighting system the normal design method is to establish the highest recommended task illuminance, then to set the average 'ambient' level at one-third of this value or at the requirement of the non-task areas (whichever is greater). The illuminance at the task area is then 'topped up' with localised or local lighting to the appropriate task level, bearing in mind that the usual uniformity requirements for the task area must be satisfied. (Calculation methods are given in the Calculations guide – see CD.)

2.3.4 Luminance and illuminance ratios

Luminance distribution in the field of view controls the adaptation level of the eyes, which affects the task visibility.

A well balanced adaptation luminance is needed to increase:

- visual acuity (sharpness of vision)
- contrast sensitivity (discrimination of small relative luminance differences)
- efficiency of ocular functions (such as accommodation, convergence, pupillary contraction, eye movements etc.).

The luminance distribution in the field of view also affects visual comfort. The following should be avoided for the reasons given:

- luminances that are too high, which may give rise to glare
- luminance contrasts that are too high, which will cause fatigue because of constant re-adaptation of the eyes
- luminance and luminance contrasts that are too low, which may result in a dull and non-stimulating working environment.

Luminance differences may be specified or measured in terms of the ratio between one luminance and another. Suggested targets are: task-to-immediate surround, 3 : 1; and task-to-general background, 10 : 1.

The reflectance and the illuminance on the surface determine the luminances of all surfaces.

Ranges of useful reflectances for the major interior surfaces are given in Table 2.2.

Table 2.2 Ranges of useful reflectances

Room surface	Reflectance range	Relative illuminance
Ceiling	0.6–0.9	0.3–0.9
Walls	0.3–0.8	0.5–0.6
Working planes	0.2–0.6	1.0
Floor	0.1–0.5	—

It has become the convention to translate luminance ratios into relative illuminances, since it is the illuminance that is used in lighting design practice. Table 2.2 gives the ranges of relative illuminances for general lighting using ceiling-mounted luminaires with a predominantly downward light distribution for a typical office. The values shown are based on research findings modified by design application and experience.

2.3.5 Room surfaces

The lighting system should play a role in reinforcing the architectural character of the interior, using daylight where possible as part of an energy-saving strategy. Spatial clarity, mood, and the visual nature of the space may be emphasised by the choice of light distribution and use of colour, as discussed in sections 1.6, Directional qualities and modelling, 1.7, Surfaces, and 3.6.3, Illuminance ratio charts.

The reflectance and finish of the major surfaces in an interior will play an important part in the use of light. High-reflectance surfaces will help inter-reflection, and are normally recommended for working interiors. This does not preclude the judicious use of colour and lower reflectance as part of the décor to give visual interest.

Matt finishes are normally recommended to avoid specular reflections or disguise surface imperfections.

2.3.5.1 Ceilings

The ceiling cavity (see Figure 3.15) will play a less significant role in a small room than it will in a large one where it can occupy a substantial part of the field of view.

The recommendation for general lighting with a predominantly downward distribution is for the ratio of average illuminance on the ceiling to the average illuminance on the horizontal working plane to be within the range 0.3–0.9.

In general the ceiling cavity reflectance should be as high as practicable, at least 0.6. The reflectance of the surface finish therefore should be of the order of 0.8 (see section 3.8.3.3, Effective reflectance).

Luminous ceilings utilising large diffusing panels are not recommended for lighting interiors for which the recommended unified glare rating is less than 19. In any case, the average luminance of such luminous ceilings should not be greater than 500 cd/m^2.

For indirect lighting, the average luminance of all surfaces forming the ceiling cavity should not be more than 500 cd/m^2. However, small areas of luminance of up to 1500 cd/m^2 will generally be acceptable, provided that sharp changes from high to low luminance are avoided (Uplighting design – see CD).

2.3.5.2 Walls

Higher reflectance of wall and partition surfaces will increase the perception of lightness in the interior. Walls with windows are a particular case. The surfaces surrounding the windows should have a reflectance of not less than 0.6 in order to reduce the contrast with the relatively bright outdoor view through the window during daytime. Windows at night form a dark specular surface, which should be covered with suitable curtains or blinds. Sharply defined patterns of light and shade ('scalloping') caused by hard-edged downlighting or wall-washing luminaires can cause a breakdown of brightness constancy, which can disrupt the visual continuity of wall surfaces.

The ratio of the average illuminance on the walls to the average illuminance on the horizontal working plane is related to the average vertical plane illuminance throughout the space. This has been shown to give good correlation with visual satisfaction for office lighting.

The recommendation is for the ratio of the average illuminance on any wall or major partition surface to the average illuminance on the horizontal working plane to be within the range 0.5–0.8.

In general, the effective reflectance of the principal walls should be between 0.3 and 0.7. The reflectance of window wall surface finishes should be at least 0.6.

2.3.5.3 Floor and working plane

The reflectance of the floor cavity plays an important role in the visual appearance of a room. With most lighting installations a proportion of the light on the ceiling will have been reflected from the floor cavity, and where this has a low reflectance it may be difficult to obtain satisfactory modelling effects without directly lighting other surfaces and so changing the luminance balance. Conversely, as the floor cavity may be one of the largest planes in a space, it is important that its luminance should not be so high as to dominate the appearance of the scene. It is therefore undesirable for the floor cavity to have an average reflectance of less than 0.20 or greater than 0.40.

The floor cavity consists of a number of surfaces: the floor, the lower parts of the walls (i.e. those below the level of the working plane), the top and sides of desks or benches, and the surfaces of other furniture or equipment. Each of these surfaces will have a particular reflectance, and its effect on the average reflectance will be in proportion to the unobscured area of the surface. It should be noted that in practical interiors it is extremely unlikely that the floor space will be unobstructed by furniture or machinery, and allowance should be made in calculating the average floor cavity reflectance (see Table 3.8). Low-reflectance bench and desk tops should generally be avoided, as these surfaces have a major influence on the average floor cavity reflectance as well as usually forming the immediate surround to the task.

In general, it is undesirable for the average floor cavity reflectance to exceed 0.40 or fall below 0.20, although it is recognised that in 'dirty' industries or heavily obstructed areas this latter figure may be difficult to achieve. In such cases, steps should be taken to avoid dark-coloured furniture and to keep working and other surfaces clean so that the average reflectance is maintained at 0.10 or above.

The reflectance of the area immediately surrounding the task should not be less than one-third of the task itself. In the case of office tasks involving white paper, this will require desktops to have a reflectance of at least 0.30.

2.3.6 Colour appearance

The 'colour appearance' of a lamp refers to the apparent colour of the light emitted, and is quantified by its correlated colour temperature (T_{CP}). (See section 1.8.1, Apparent colour of emitted light, for more information on the colour appearance of light sources.)

The choice of colour appearance is a matter of psychology, aesthetics, and what is considered natural. The following general rules may help with the selection of light source colour:

— for rooms lit to an illuminance of 300 lux or less, a warm or intermediate colour is preferred; cold apparent colour lamps tend to give rooms a gloomy appearance at lower illuminances
— where it is desirable to blend with daylight, intermediate correlated colour temperature (CCT) sources should be used
— different colour lamps should not be used haphazardly in the same room.

However, where light sources with good colour rendering are used, there is no evidence of a simple relationship between CCT and people's preference of the space.

2.3.7 Colour rendering

It is important for visual performance and the feeling of comfort and well being that colours in the environment, of objects and of human skin are rendered naturally, correctly, and in a way that makes people look attractive and healthy.

Safety colours according to *ISO 3864* must always be recognisable as such.

To provide an objective indication of the colour rendering properties of a light source, the general colour-rendering index, R_a, has been introduced. The maximum value of R_a is 100, and this figure decreases with decreasing colour-rendering quality. Lamps with a colour-rendering index lower than 80 should not be used in interiors where people work or stay for longer periods. Exceptions may apply for some places and/or activities (e.g. high-bay lighting), but suitable measures shall be taken to ensure lighting with higher colour rendering at fixed continually occupied work places and where safety colours have to be recognised. The minimum values of colour-rendering index for distinct types of interiors (areas), tasks or activities are given in section 2.5, Lighting schedule.

2.3.8 Modelling and emphasis

Directional lighting may be used to highlight objects, reveal texture, and improve the appearance of people within the space. This is described by the term 'modelling'. Directional lighting of a visual task may also affect its visibility.

Modelling is the balance between diffuse and directional light, and is a valid criterion of lighting quality in virtually all types of interiors. The general appearance of an interior is enhanced when its structural features and the people and objects within it are lit so that form and texture are revealed clearly and pleasantly. This occurs when the light comes predominantly from one direction; the shadows so essential to good modelling are then formed without confusion.

The lighting should not be too directional or it will produce harsh shadows; neither should it be too diffuse or the modelling effect will be lost entirely, resulting in a very dull luminous environment.

The relationship between the intensity of the directional lighting and the diffuse illuminance is expressed as the vector/scalar ratio (see section 3.6.3, Illuminance ratio charts). This objective ratio is a useful criterion when considering the relative values of directional lighting to non-directional or reflected lighting. A vector/scalar ratio from 1.2 to 1.8 will prove satisfactory in normal general lighting conditions where perception of faces is important. Under such conditions, facial modelling will usually appear balanced and natural.

Display lighting calls for greater impact and emphasis. Table 2.3 is intended to give general guidance on the display illuminance ratios (DIR) that must be provided to achieve increasing degrees of emphasis from 'subtle' to 'dramatic' (Figure 2.1). The display illuminance ratio is that between the general horizontal plane illuminance in the room and the value of local illuminance in the plane of the object to be displayed. Greater degrees of emphasis are likely to require lower values of general diffused lighting to avoid the need for excessive values of local display illuminance.

Table 2.3 Direct illuminance ratio

Display effect	Objective display illuminance ratio (DIR)	Subjective apparent brightness ratio
Subtle	5 : 1	2.5 : 1
Moderate	15 : 1	5 : 1
Strong	30 : 1	7 : 1
Dramatic	50 : 1	10 : 1

Table 2.3 also shows that, because of visual adaptation, the apparent brightness difference between the object on display and its surroundings is less than the measured illuminance difference.

2.3.9 Glare

Glare is the sensation produced by bright areas in the field of view, and may be experienced either as discomfort glare or as disability glare (see section 1.5, Glare, for more information).

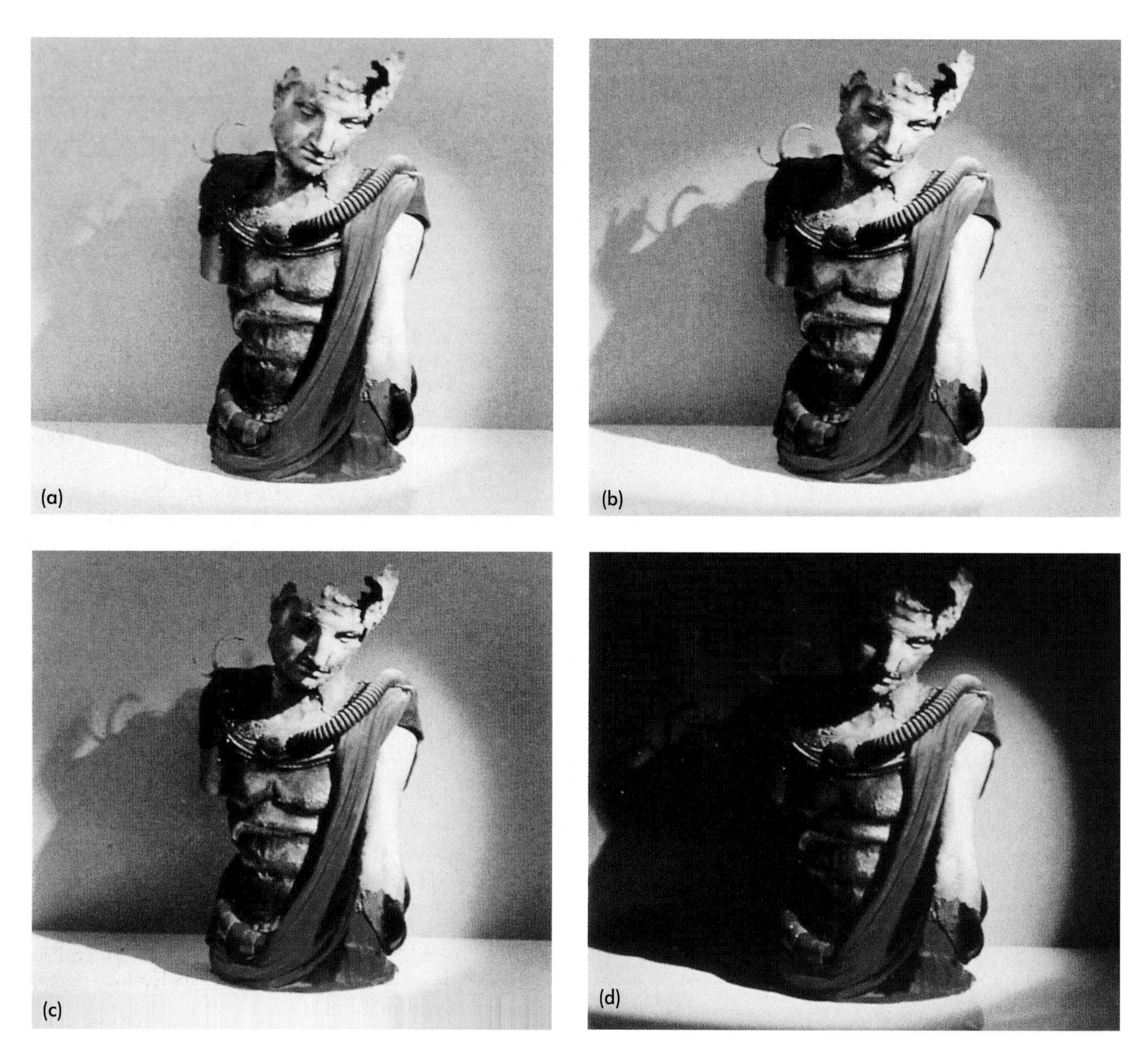

Figure 2.1 Display illuminance ratio (DIR): (a) 'subtle', with a DIR of approximately 5 : 1; (b) 'moderate', with a DIR of 15 : 1; (c) 'strong', with a DIR of 30 : 1; (d) 'dramatic', with a DIR of 50 : 1

In order to control the discomfort glare in installations, the Lighting schedule (section 2.5) gives limits for the value of UGR. All assumptions made in the determination of UGR must be stated in the scheme documentation.

2.3.9.1 Shielding against glare

Bright light sources can cause glare and impair the vision of objects. Suitable shielding of lamps and the shading of windows with blinds can eliminate this effect (see section 1.5, Glare).

2.3.10 Lighting of work stations with display screen equipment

The lighting for the DSE (display screen equipment) work stations must be appropriate for all the tasks performed at the work station, e.g. reading from the screen or printed text, writing on paper, and keyboard work.

For these areas, the lighting criteria and system shall be chosen in accordance with the activity area, task type and type of interior, as set out in the Lighting schedule (section 2.5).

The DSE and, in some circumstances, the keyboard may suffer from reflections, causing disability and discomfort glare. It is therefore necessary to select, locate and arrange the luminaires to avoid high brightness reflections.

2.3.10.1 Luminaire luminance limits

Table 2.4 gives the limits of the average luminaire luminance at elevation angles of 65° and above from the downward vertical, and radially around the luminaires, for work places where the screens (which are vertical or inclined up to a 15° tilt angle) are used.

Note: For certain special places using, for example, sensitive screens or variable inclination, the luminance limits in the table should be applied for a lower elevation angle (e.g. 55°) of the luminaire.

Table 2.4 VDT screen class and maximum luminaire luminance

Screen classes in accordance with BS EN *ISO 9241-7*	Maximum luminance (cd/m^{-2}) where some negative polarity used
Type I and II, good or moderate screen treatment	1000
Type III, no screen treatment	200

Where positive polarity software only is used on Type I or II screens, the luminance limit can be increased to 1500 cd/m^{-2}; where positive polarity software only is used on Type III screens, the luminance limit can be increased to 500 cd/m^{-2}.

2.4 Energy efficiency recommendations

Lighting must provide a suitable visual environment within a particular space – sufficient and suitable lighting for the performance of a range of tasks, provision of a desired appearance etc. This objective should be achieved without waste of energy. However, it is important not to compromise the visual aspects of a lighting installation simply to reduce energy consumption. In most organisations the cost of lighting energy, although substantial, is only a small fraction of the total costs associated with the activity in the space. For example, the impact of poor visual conditions on work quality and productivity costs is likely to be many times greater than the lighting energy costs in an office or in a factory (labour costs may typically be around 100 times greater than lighting energy costs). Similarly, in a shop the sales turnover resulting from correct display of merchandise will be very much greater than the energy costs for lighting. It is thus a false economy to save energy at the expense of human effectiveness.

On the other hand, the profligate and unnecessary use of energy and the associated costs, both financial and environmental, should be avoided. The environmental impact of lamps is discussed in the Lamps and the environment section of the LIF Lamp Guide (see CD).

The recommendations that follow provide guidance on energy efficiency for lighting installations. They assume that good design has been combined with the use of modern equipment. New lighting designs should normally meet these levels. The recommendations can also be used to gauge the efficiency of existing installations and to determine whether or not the existing installation needs remedial action to achieve acceptable energy efficiency.

The subject of energy use is also covered in the Building Regulations Part L. There are two approved documents that are in force from April 2002. They are:

- L1 – Conservation of fuel power in dwellings
- L2 – Conservation of fuel and power in buildings other than dwellings.

Part J of the Building Standards (Scotland) Regulations include similar requirements. The following three sections give further guidance on this topic.

2.4.1 Power and time

The energy (kW/h) used by a lighting installation depends on both the power (kW) and the time (h). Energy efficiency can be achieved:

- by using the most efficient lighting equipment (see CD – Lighting Equipment) to obtain the desired lighting solution, i.e. the electrical load (kW) is kept to a minimum while achieving the lighting design objectives
- by using effective controls so that the lighting is not in operation at times when it is not needed, i.e. the period of operation (h) is kept to a minimum.

The lighting designer can limit the electrical power loading and the use of energy, but it is the operator of the installation who will ultimately be responsible for achieving high energy efficiency in practice.

2.4.2 Energy efficient equipment

Information on the energy efficiency of lamps and luminaires is given in Lighting equipment (see CD). While the lighting requirements for different spaces within a building can be met most appropriately using different lamps or luminaires, an average initial circuit lamp luminous efficacy of at least 65 lm/W for the fixed lighting equipment within the building should be achieved. Both emergency lighting systems and equipment that is not fixed (e.g. track-mounted luminaires) are excluded from this figure. Thus it is possible to use equipment of lower energy efficiency (e.g. tungsten-halogen spotlights) in some areas, combined with more energy efficient equipment (e.g. fluorescent lamps with high frequency electronic ballasts) in other areas. This recommendation can be used as a guideline at the design concept stage, but it does not take account of energy use.

In practice, much energy is wasted outside normal working hours by lighting being left on when not required, although some lighting may be needed for cleaning and security. Override controls should provide full lighting in emergency conditions at night. Similarly, lighting may not be needed during working hours if there is sufficient daylight or if spaces in the building are vacant. Adequate lighting controls should be installed to allow the building occupants to use only that lighting which is actually needed at any particular time. The control system should be flexible enough to allow an appropriate level of lighting to be achieved and lighting that is not required to be switched off. This may be achieved by:

- localised switching, using switches provided throughout the space and not concentrated at the point of entry
- time switching, providing automatic switching of luminaires to a predetermined schedule
- automatic switching or dimming of lighting in relation to occupancy and daylight level measured by a photoelectric sensor.

Further details of such control systems are given in sections 3.7, Energy management, and 3.8.1, Costs and energy use; also in Lighting controls (see CD).

The ultimate aim must be to achieve the desired lighting solution at the lowest practical energy use. It is possible that a higher installed load combined with a suitable control system to give low hours of use will result in lower energy consumption than an alternative installation with a lower power loading but poorer control. It is thus important to consider both aspects.

2.4.3 Lighting energy targets

Table 2.5 provides targets of power density, averaged over the space, for general lighting for a range of applications with particular task illuminance values. The values of average installed power density per application are based on current good practice using efficient lamps and luminaires in good quality installations; however, improvements on these values could be possible. The targets are for an average-sized space (room index of 2.5), with high room surface reflectances (ceiling reflectance 0.7, wall reflectance 0.5, and floor reflectance 0.2) and a high degree of installation maintenance. The values are in average watts/metre^2 for the space and at particular task illuminance levels.

Table 2.5 Lighting energy targets

Lamp type	CIE general colour-rendering index (R_a)	Task illuminance (lux)	Average installed power density (W/m^2)
Commercial and other similar application, e.g. offices, shops and schools*			
Fluorescent – triphosphor	80–90	300	7
		500	11
		750	17
Compact fluorescent	80–90	300	8
		500	14
		750	21
Metal halide	60–90	300	11
		500	18
		750	27
Industrial and manufacturing applications			
Fluorescent – triphosphor	80–90	300	6
		500	10
		750	14
		1000	19
Metal halide	60–90	300	7
		500	12
		750	17
		1000	23
High pressure sodium	40–80	300	6
		500	11
		750	16
		1000	21

*Values do not include energy for display lighting.

2.5 Lighting schedule

The lighting schedule gives recommendations for the lighting of various areas in terms of the following parameters.

Maintained illuminance (lux): the maintained illuminance in lux for the area. This value may be adjusted to suit a particular task (see section 2.3.2, Illuminance). The uniformity of the illuminance is given in section 2.3.3, Illuminance variation. For more information about maintenance factors, see section 3.5.2, Maintained illuminance.

Limiting glare rating: limiting glare rating is the maximum discomfort glare, expressed as UGR, permitted for a given application. To calculate UGR for a given installation by using tables, see CD (Sample glare rating calculation using a standard table).

Minimum colour rendering (R_a): this value is the minimum colour rendering value of the lamps use in the installation (see section 2.3.7, Colour rendering).

The schedule gives recommendations for the lighting for the following types of area:

Agriculture

	Maintained illuminance (lux)	Limiting glare rating	Minimum colour rendering (R_a)	Notes
Loading and operating goods handling equipment and machinery	200	25	80	
Buildings for livestock	50	–	40	
Sick animal pens; calving stalls	200	25	80	1
Feed preparation; dairy; utensil washing	200	25	80	2

Notes:

1. A lower illuminance is acceptable in the absence of the stockman
2. Luminaires suitable for being hosed down may be required in some areas

Illuminance values may be varied to suit circumstances; see section 2.3.2, Illuminance

Airports

	Maintained illuminance (lux)	Limiting glare rating	Minimum colour rendering (R_a)	Notes
Arrival and departure halls, baggage claim areas	200	22	80	1
Connecting areas, escalators, travolators	150	22	80	
Information desks, check-in desks	500	19	80	2
Customs and passport control desks	500	19	80	3
Waiting areas	200	22	80	
Luggage store rooms	200	25	80	
Security check areas	300	19	80	2
Air traffic control tower	500	16	80	2, 4, 5, 6
Testing and repair hangers	500	22	80	1
Engine test areas	500	22	80	1
Measuring areas in hangers	500	22	80	1

Notes:

1. If high-bay lighting is used the colour rendering requirement may be relaxed, provided measures are taken to ensure lighting with higher colour rendering is provided at continually occupied work stations.
2. See section 2.3.10, Lighting of work stations with display screen equipment.
3. Vertical illuminance is important.
4. Lighting should be dimmable.
5. Glare from daylight should be avoided.
6. Avoid reflections in windows, especially at night.

Illuminance values may be varied to suit circumstances; see section 2.3.2, Illuminance.

Bakeries

	Maintained illuminance (lux)	Limiting glare rating	Minimum colour rendering (R_a)
Preparation and baking	300	22	80
Finishing, glazing, decorating	500	22	80

Illuminance values may be varied to suit circumstances; see section 2.3.2, Illuminance.

Cement, cement goods, concrete and bricks

	Maintained illuminance (lux)	Limiting glare rating	Minimum colour rendering (R_a)	Notes
Dyeing	50	28	20	1
Preparation work on mixers and kilns	200	28	40	
General machine work	300	25	80	2
Rough forms	300	25	80	2

Notes:

General: lamps and luminaires may be subject to severe fouling and high ambient temperatures. Dustproof (IP5X) or other luminaires with good maintenance properties are desirable in areas where raw materials are formed into the basic product.

1. Safety colours should be recognisable.
2. If high-bay lighting is used the colour rendering requirement may be relaxed, provided measures are taken to ensure lighting with higher colour rendering is provided at continually occupied work stations.

Illuminance values may be varied to suit circumstances; see section 2.3.2, Illuminance.

Ceramic, tiles and glassware

	Maintained illuminance (lux)	Limiting glare rating	Minimum colour rendering (R_a)	Notes
Dyeing	50	28	20	1
Preparation, general machine work	300	25	80	2
Enamelling, rolling, pressing, shaping simple parts, glazing, glass blowing	300	25	80	2
Grinding, engraving, glass polishing, shaping precision parts, manufacture of glass instruments	750	19	80	2
Grinding of optical glass, crystal, hand grinding and engraving	750	16	80	
Precision work, e.g. decorative grinding, hand painting	1000	16	90	3
Manufacture of synthetic precious stones	1500	16	90	3

Notes:

General: lamps and luminaires may be subject to severe fouling and high ambient temperatures. Dustproof (IP5X) or other luminaires with good maintenance properties are desirable in areas where raw materials are formed into the basic product.

1. Safety colours should be recognisable.
2. If high-bay lighting is used the colour rendering requirement may be relaxed, provided measures are taken to ensure lighting with higher colour rendering is provided at continually occupied work stations.
3. Colour temperature of the light should be greater than 4000 K.

Illuminance values may be varied to suit circumstances; see section 2.3.2, Illuminance.

Chemical, plastics and rubber industry

	Maintained illuminance (lux)	Limiting glare rating	Minimum colour rendering (R_a)	Notes
Remote-operated processing installations	50	–	20	1, 2
Processing with limited manual intervention	150	28	40	
Constantly manned work places in processing installations	300	25	80	
Precision measuring rooms, laboratories	500	19	80	
Pharmaceutical production	500	22	80	
Tyre production	500	22	80	
Colour inspection	1000	16	90	3
Cutting, finishing and inspection	750	19	80	

Notes:

1. Safety colours should be recognisable.
2. Supplementary local lighting may be needed for maintenance work.
3. Colour temperature of the light should be greater than 4000 K.

Illuminance values may be varied to suit circumstances; see section 2.3.2, Illuminance.

Control rooms

	Maintained illuminance (lux)	Limiting glare rating	Minimum colour rendering (R_a)	Notes
Plant rooms, switch gear rooms	200	25	60	1
Telex, post room, switchboard	500	19	80	2

Notes:

1. Localised lighting of control display and control desks may be appropriate. Care should be taken to avoid shadows and veiling reflections on the instruments and VDT screens.
2. In switchboard areas, avoid veiling reflections from controls. Too high an illuminance may reduce the visibility of signal lights. Supplementary local lighting may be desirable where directories are used.

Illuminance values may be varied to suit circumstances; see section 2.3.2, Illuminance.

Educational buildings

	Maintained illuminance (lux)	Limiting glare rating	Minimum colour rendering (R_a)	Notes
Classrooms, tutorial rooms	300	19	80	1
Classrooms for evening classes and adult education	500	19	80	1
Lecture hall	500	19	80	1
Blackboard	500	19	80	2
Demonstration table	500	19	80	3
Art rooms	500	19	80	
Art rooms in art schools	750	19	90	4
Technical drawing rooms	750	16	80	
Practical rooms and laboratories	500	19	80	
Handicraft rooms	500	19	80	
Teaching workshops	500	19	80	
Music practice rooms	300	19	80	
Computer practice rooms	300	19	80	5
Language laboratory	300	19	80	
Preparation rooms and workshops	500	22	80	
Entrance halls	200	22	80	
Circulation areas, corridors	100	25	80	
Stairs	150	25	80	
Student common rooms and assembly halls	200	22	80	
Teachers' rooms	300	19	80	
Stock rooms for teaching materials	100	25	80	
Sports halls, gymnasiums, swimming pools	300	22	80	6
School canteens	200	22	80	
Kitchen	500	22	80	

Notes:

1. Lighting should be controllable.
2. Prevent specular reflections.
3. In lecture halls, maintained illuminance should be 750 lux.
4. The colour temperature of the light should be greater than 5000 K.
5. See section 2.3.10, Lighting of work stations with display screen equipment.
6. See CIBSE *Lighting Guide 4: Sports*.

Illuminance values may be varied to suit circumstances; see section 2.3.2, Illuminance.

Electrical industry

	Maintained illuminance (lux)	Limiting glare rating	Minimum colour rendering (R_a)	Notes
Cable and wire manufacture	300	25	80	1, 2
Winding:				
— large coils	300	25	80	1, 2
— medium-sized coils	500	22	80	1
— small coils	750	19	80	1
Coil impregnating	300	25	80	1, 2
Galvanising	300	25	80	1, 2
Assembly work:				
— rough (e.g. large transformers)	300	25	80	1, 2
— medium (e.g. switchboards)	500	22	80	1, 3
— fine (e.g. telephones)	750	19	80	1, 3
— precision (e.g. measuring equipment)	1000	16	80	1, 3
Electronic workshops, testing, adjusting	1500	16	80	3
Printed circuit boards:				
— printing	500	22	80	
— hand insertion of components	750	19	80	3, 4
— soldering	750	19	80	3, 4
— inspection	1000	16	80	3, 4, 5

Notes:

1. If high-bay lighting is used the colour rendering requirement may be relaxed, provided that measures are taken to ensure lighting with higher colour rendering is provided at continually occupied work stations.
2. With large machines some obstruction is likely, and portable or local lighting may be needed.
3. Local lighting may be appropriate.
4. UV/blue visible filtering sleeves (280–450 nm) may be required where certain types of sensitive components are to be processed.
5. A large, low luminance overhead luminaire ensures specular reflection conditions that are helpful for the inspection of printed circuit boards.

Illuminance values may be varied to suit circumstances; see section 2.3.2, Illuminance.

Food stuffs and luxury food industry

	Maintained illuminance (lux)	Limiting glare rating	Minimum colour rendering (R_a)	Notes
Work places and zones:				
— in breweries, malting floor, for washing, barrel filling, cleaning, sieving, peeling	200	25	80	
— for cooking in preserve and chocolate factories	–	–	–	
— for drying and fermenting raw tobacco, fermentation cellar	–	–	–	
Sorting and washing of products, milling mixing, packing	300	25	80	
Workplaces and critical zones in slaughter houses, dairies, mills, on filtering floor in sugar refineries	500	25	80	1, 2
Cutting and sorting of fruit and vegetables	300	25	80	
Manufacture of delicatessen foods, kitchen work, manufacture of cigars and cigarettes	500	22	80	
Inspection of glasses and bottles, product control, trimming, sorting, decoration	500	22	80	
Laboratories	500	19	80	
Colour inspection	1000	16	90	3

Notes:

General: luminaires should be constructed so that no part of the luminaire can fall into the foodstuffs, even when the luminaire is opened for lamp changing. The luminaires should be capable of being washed or hosed down in safety. Lamps suitable for operation at low temperatures will be necessary for some food storage areas. Lamps and luminaires suitable for hot and humid conditions may be required for some other areas.

1. Areas containing a dust explosion hazard may be present; appropriate luminaires should be chosen.
2. Damp conditions may be present and hosing down may be part of the cleaning process. For meat inspection there is a statutory minimum illuminance of 540 lux (50 lumens/ft^2), and lamps with a colour rendering index greater than 90 should be used.
3. Colour temperature of the light should be greater than 4000 K.

Illuminance values may be varied to suit circumstances, see section 2.3.2, Illuminance.

Foundries and metal casting

	Maintained illuminance (lux)	Limiting glare rating	Minimum colour rendering (R_a)	Notes
Man-sized under floor tunnels, cellars etc.	50	–	20	1
Platforms	100	25	40	
Sand preparation	200	25	80	2, 3
Dressing room	200	25	80	2
Work places at cupola and mixer	200	25	80	2
Casting bay	200	25	80	2
Shake out areas	200	25	80	2, 3
Machine moulding	200	25	80	2, 3
Hand and core moulding	300	25	80	2, 4
Die casting	300	25	80	2
Model building	500	25	80	2, 4

Notes:

General: Lamps and luminaires may be subject to severe fouling and high ambient temperatures. Dustproof (IP5X) or other luminaires with good maintenance properties are desirable in areas where raw materials are formed into the basic product.

1. Safety colours should be recognisable.
2. If high-bay lighting is used the colour rendering requirement may be relaxed, provided that measures are taken to ensure lighting with higher colour rendering is provided at continually occupied work stations.
3. If blast cleaning is used, luminaires should be positioned away from the work area. Where metal castings are cleaned by means of abrasive wheels or bands, then the dust produced may present an explosion hazard; luminaires should be chosen appropriately.
4. Light distribution needs to be diffused and flexible to ensure good lighting of deep moulds.

Illuminance values may be varied to suit circumstances; see section 2.3.2, Illuminance.

Hairdressers

	Maintained illuminance (lux)	Limiting glare rating	Minimum colour rendering (R_a)
Hairdressing	500	19	90

Illuminance values may be varied to suit circumstances; see section 2.3.2, Illuminance.

Healthcare – autopsy rooms and mortuaries

	Maintained illuminance (lux)	Limiting glare rating	Minimum colour rendering (R_a)	Notes
General lighting	500	19	90	1
Autopsy table and dissecting table	5000	–	90	2, 3

Notes:

1. Luminaires may be subject to an aggressive cleaning regime.
2. Examination luminaire is required.
3. Values greater than 5000 lux may be required.

Illuminance values may be varied to suit circumstances; see section 2.3.2, Illuminance.

Healthcare – decontamination rooms

	Maintained illuminance (lux)	Limiting glare rating	Minimum colour rendering (R_a)	Note
Sterilisation rooms	300	22	80	1
Disinfection rooms	300	22	80	1

Note:

1. Luminaires may be subject to high humidity and temperatures as well as an aggressive cleaning regime.

Illuminance values may be varied to suit circumstances; see section 2.3.2, Illuminance.

Healthcare – delivery rooms

	Maintained illuminance (lux)	Limiting glare rating	Minimum colour rendering (R_a)	Note
General lighting	300	19	80	
Examination and treatment	1000	19	80	1

Note:

1. Examination luminaire may be required.

Illuminance values may be varied to suit circumstances; see section 2.3.2, Illuminance.

Healthcare – dentists

	Maintained illuminance (lux)	Limiting glare rating	Minimum colour rendering (R_a)	Notes
General lighting	500	19	90	1
At the patient	1000	–	90	
Operating cavity	5000	–	90	2, 3
White teeth matching	5000	–	90	4

Notes:

1. Lighting should be glare free for the patient.
2. Values greater than 5000 lux may be needed.
3. A dental inspection luminaire is required.
4. Colour temperature of the light should be greater than 6000 K.

Illuminance values may be varied to suit circumstances; see section 2.3.2, Illuminance

Healthcare – ear examination rooms

	Maintained illuminance (lux)	Limiting glare rating	Minimum colour rendering (R_a)	Note
General lighting	300	19	80	
Ear examination	1000	–	90	1

Note:

1. Examination luminaire may be required.

Illuminance values may be varied to suit circumstances; see section 2.3.2, Illuminance.

Healthcare – examination rooms

	Maintained illuminance (lux)	Limiting glare rating	Minimum colour rendering (R_a)	Note
General lighting	500	19	90	
Examination and treatment	1000	19	90	1

Note:

1. Examination luminaire may be required.

Illuminance values may be varied to suit circumstances; see section 2.3.2, Illuminance.

Healthcare – eye examination rooms

	Maintained illuminance (lux)	Limiting glare rating	Minimum colour rendering (R_a)	Notes
General lighting	300	19	80	
Examination of the outer eye	1000	–	90	1
Reading and colour vision test with vision charts	500	16	90	1

Note:

1. Examination luminaire may be required.

Illuminance values may be varied to suit circumstances; see section 2.3.2, Illuminance.

Healthcare – intensive care

	Maintained illuminance (lux)	Limiting glare rating	Minimum colour rendering (R_a)	Notes
General lighting	100	19	90	1
Simple examinations	300	19	90	2
Examination and treatment	1000	19	90	2
Night watch	20	19	90	1

Notes:

1. Illuminance at floor level.
2. Illuminance at bed level.

Illuminance values may be varied to suit circumstances; see section 2.3.2, Illuminance.

Healthcare – laboratories and pharmacies

	Maintained illuminance (lux)	Limiting glare rating	Minimum colour rendering (R_a)	Note
General lighting	500	19	80	
Colour inspection	1000	19	90	1

Note:

1. Colour temperature of the light should be greater than 6000 K.

Illuminance values may be varied to suit circumstances; see section 2.3.2, Illuminance.

Healthcare – operating areas

	Maintained illuminance (lux)	Limiting glare rating	Minimum colour rendering (R_a)	Note
Pre-op and recovery rooms	500	19	90	
Operating theatre	1000	19	90	
Operating cavity	–	–	–	1

Note:

1. Operating luminaire required; illuminance 10 000–100 000 lux.

Illuminance values may be varied to suit circumstances; see section 2.3.2, Illuminance.

Healthcare – scanner rooms

	Maintained illuminance (lux)	Limiting glare rating	Minimum colour rendering (R_a)	Note
General lighting	300	19	80	
Scanners with image enhancers and television systems	50	19	80	1

Note:

1. See section 2.3.10, Lighting of work stations with display screen equipment.

Illuminance values may be varied to suit circumstances; see section 2.3.2, Illuminance.

Healthcare – treatment rooms

	Maintained illuminance (lux)	Limiting glare rating	Minimum colour rendering (R_a)	Note
Dialysis	500	19	80	1
Dermatology	500	19	90	
Endoscope rooms	300	19	80	
Plaster rooms	500	19	80	
Medical baths	300	19	80	
Massage and radiotherapy	300	19	80	

Note:

1. The lighting should be controllable.

Illuminance values may be varied to suit circumstances; see section 2.3.2, Illuminance.

Healthcare – wards

	Maintained illuminance (lux)	Limiting glare rating	Minimum colour rendering (R_a)	Note
General lighting	100	19	80	1, 2
Reading lighting	300	19	80	2, 3
Simple examinations	300	19	80	2
Examination and treatment	1000	19	80	
Night lighting, observation lighting	5	–	80	2
Bathrooms and toilets for patients	200	22	80	

Notes:

1. Illuminance at floor level.
2. Prevent high luminances in the patient's field of view.
3. Local lighting required.

Illuminance values may be varied to suit circumstances; see section 2.3.2, Illuminance.

Healthcare premises – general rooms

	Maintained illuminance (lux)	Limiting glare rating	Minimum colour rendering (R_a)	Note
Waiting rooms	200	22	80	1
Corridors: during the day	200	22	80	1
Corridors: at night	50	22	80	1
Day rooms	200	22	80	1
Staff office	500	19	80	
Staff rooms	300	19	80	

Note:

1. Illuminance at floor level.

Illuminance values may be varied to suit circumstances; see section 2.3.2, Illuminance.

Hotels and restaurants

	Maintained illuminance (lux)	Limiting glare rating	Minimum colour rendering (R_a)	Notes
Reception/cashier desk, porters desk	300	22	80	1
Kitchen	500	22	80	2
Restaurant, dining room, function room	–	–	80	3
Self-service restaurant	200	22	80	
Buffet	300	22	80	
Conference rooms	500	19	80	4
Corridors	100	25	80	5

Notes:

1. Localised lighting may be appropriate.
2. There should be a transition zone between kitchen and restaurant.
3. The lighting should be designed to create the appropriate atmosphere.
4. Lighting should be controllable.
5. During the night lower levels may be acceptable. Late-night low-level lighting may be used with manual override or presence detection.

Illuminance values may be varied to suit circumstances; see section 2.3.2, Illuminance.

Jewellery manufacture

	Maintained illuminance (lux)	Limiting glare rating	Minimum colour rendering (R_a)	Note
Working with precious stones	1500	16	90	1
Manufacture of jewellery	1000	16	90	
Watch making (manual)	1000	16	80	
Watch making (automatic)	500	19	80	

Note:

1. Colour temperature to be at least 4000 K.

Illuminance values may be varied to suit circumstances; see section 2.3.2, Illuminance.

Laundries

	Maintained illuminance (lux)	Limiting glare rating	Minimum colour rendering (R_a)
Goods in, marking and sorting	300	25	80
Washing and dry cleaning	300	25	80
Ironing, pressing	300	25	80
Inspection and repairs	750	19	80

Note:

General: luminaires may be subject to a warm, humid atmosphere.

Illuminance values may be varied to suit circumstances; see section 2.3.2, Illuminance.

Leather and leather goods

	Maintained illuminance (lux)	Limiting glare rating	Minimum colour rendering (R_a)	Note
Work on vats, barrels, pits	200	25	40	
Fleshing, skiving, rubbing, tumbling of skins	300	25	80	
Saddlery work, shoe manufacture: stitching, sewing, polishing, shaping, cutting, punching	500	22	80	
Sorting	500	22	90	1
Leather dyeing (machine)	500	22	80	
Quality control	1000	19	80	
Colour inspection	1000	16	90	1
Shoe making	500	22	80	
Glove making	500	22	80	

Note:

1. Colour temperature of the light should be greater than 4000 K.

Illuminance values may be varied to suit circumstances; see section 2.3.2, Illuminance.

Libraries

	Maintained illuminance (lux)	Limiting glare rating	Minimum colour rendering (R_a)	Notes
Bookshelves	200	19	80	1
Reading areas	500	19	80	2
Counters	500	19	80	2

Notes:

1. The illuminance should be provided on the vertical face at the bottom of the bookshelf.
2. Local or localised lighting may be appropriate.

Illuminance values may be varied to suit circumstances; see section 2.3.2, Illuminance.

Lighting recommendations for traffic zones

	Maintained illuminance (lux)	Limiting glare rating	Minimum colour rendering (R_a)	Notes
Circulation areas and corridors	100	28	40	1, 2, 3, 4
Stairs, escalators, travolators	150	25	40	2, 5, 6, 7
Loading ramps/bays	150	25	40	8

Notes:

1. Illuminance to be measured at floor level.
2. The R_a and glare to be similar to adjacent areas, the lighting of exits and entrances to provide a transition zone to avoid sudden changes in direction.
3. The illuminance to be increased to 150 lux if there are vehicles on the route; this includes all corridors in hospitals.
4. In residential areas the illuminance may be reduced to 20 lux. Late-night low-level lighting may be used with manual override or presence detection.
5. Stairs should be lit to provide a contrast between the treads and the risers. Avoid specular reflections on the treads. Safety will be enhanced by the use of coloured nosings that contrast with the finish of the treads and risers.

6. For escalators, below step lighting may be effective in providing contrast between the steps and risers.
7. Increased illuminance may be necessary at the entrances and exits of escalators and travolators.
8. Avoid glare to drivers of vehicles approaching the loading bay. Light and clearly mark the edge of the loading bay.

Illuminance values may be varied to suit circumstances; see section 2.3.2, Illuminance.

Metal working and processing

	Maintained illuminance (lux)	Limiting glare rating	Minimum colour rendering (R_a)	Notes
Open die forging	200	25	60	
Drop forging	300	25	60	
Welding	300	25	60	1
Rough and average machining: tolerances ≥ 0.1 mm	300	22	60	2
Precision machining and grinding: tolerances < 0.1 mm	500	19	60	2
Scribing, inspection	750	19	60	3
Wire and pipe drawing shops; cold forming	300	25	60	
Plate machining: thickness ≥ 5 mm	200	25	60	2
Sheet metalwork: thickness < 5 mm	300	22	60	2
Tool making: cutting equipment manufacture	750	19	60	
Assembly:				
— rough	200	25	80	2, 4
— medium	300	25	80	2, 4
— fine	500	22	80	2, 4
— precision	750	19	80	4
Galvanising	300	25	80	4
Surface preparation and painting	750	25	80	4
Tool, template and jig making, precision mechanics, micro-mechanics	1000	19	80	

Notes:

1. Care is necessary to prevent exposure of eyes and skin to radiation. Welding screens will be used, so considerable obstruction is likely. Portable lighting may be useful.
2. Some obstruction is likely. Care should be taken to minimise stroboscopic effects on rotating machinery.
3. Care should be taken to avoid multiple shadows.
4. If high-bay lighting is used the colour rendering requirement may be relaxed, provided that measures are taken to ensure lighting with higher colour rendering is provided at continually occupied work stations.

Illuminance values may be varied to suit circumstances; see section 2.3.2, Illuminance.

Nursery and play schools

	Maintained illuminance (lux)	Limiting glare rating	Minimum colour rendering (R_a)
Play room	300	19	80
Nursery	300	19	80
Handicraft room	300	19	80

Illuminance values may be varied to suit circumstances; see section 2.3.2, Illuminance.

Offices

	Maintained illuminance (lux)	Limiting glare rating	Minimum colour rendering (R_a)	Notes
Filing, copying etc.	300	19	80	1
Writing, typing, reading, data processing	500	19	80	2
Technical drawing	750	16	80	
CAD work stations	500	19	80	2
Conference and meeting rooms	500	19	80	3
Reception desk	300	22	80	
Archives	200	25	80	1

Notes:

1. For filing, the vertical surfaces are especially important.
2. See section 2.3.10, Lighting of work stations with display screen equipment.
3. The lighting should be controllable.

Illuminance values may be varied to suit circumstances; see section 2.3.2, Illuminance.

Paper and paper goods

	Maintained illuminance (lux)	Limiting glare rating	Minimum colour rendering (R_a)	Note
Edge runners, pulp mills	200	25	80	1
Paper manufacture and processing, paper and corrugating machines, cardboard manufacture	300	25	80	1
Standard bookbinding work, e.g. folding, sorting, gluing, cutting, embossing, sewing	500	22	80	

Note:

General: luminaires may be subject to a warm, humid atmosphere.

1. If high-bay lighting is used the colour rendering requirement may be relaxed, provided that measures are taken to ensure lighting with higher colour rendering is provided at continually occupied work stations.

Illuminance values may be varied to suit circumstances; see section 2.3.2, Illuminance.

Places of public assembly – general areas

	Maintained illuminance (lux)	Limiting glare rating	Minimum colour rendering (R_a)	Note
Entrance halls	100	22	80	1
Cloakrooms	200	25	80	
Lounges	200	22	80	
Ticket offices	300	22	80	

Note:

1. UGR only if applicable.

Illuminance values may be varied to suit circumstances; see section 2.3.2, Illuminance.

Power stations

	Maintained illuminance (lux)	Limiting glare rating	Minimum colour rendering (R_a)	Notes
Fuel supply plant	50	–	20	1
Boiler house	100	28	40	1
Machine halls	200	25	80	2, 3
Side rooms, e.g. pump rooms, condenser rooms etc., switchboards (inside buildings)	200	25	60	4
Control rooms	500	16	80	5
Outdoor switch gear	20	–	20	1

Notes:

1. Safety colours should be recognisable.
2. If high-bay lighting is used the colour rendering requirement may be relaxed, provided that measures are taken to ensure lighting with higher colour rendering is provided at continually occupied work stations.
3. Additional local lighting of instruments and inspection points may be required.
4. In areas such as ash handling plants, settling pits and battery rooms there may be corrosive and hazardous atmospheres.
5. Control panels are often vertical and dimming may be required. VDTs may be used. It may be necessary to use luminaires with limited luminance at high angles.

Illuminance values may be varied to suit circumstances; see section 2.3.2, Illuminance.

Printers

	Maintained illuminance (lux)	Limiting glare rating	Minimum colour rendering (R_a)	Notes
Cutting, gilding, embossing, block engraving, work on stones and platens, printing machines, matrix making	500	19	80	
Paper sorting and hand printing	500	19	80	
Type setting, retouching, lithography	1000	19	80	1, 2
Colour inspection in multicoloured printing	1500	16	90	3
Steel and copper engraving	2000	16	80	4

Notes:

1. Large area, low-luminance luminaires are desirable.
2. Local lighting may be appropriate.
3. Colour temperature of the light should be greater than 4000 K.
4. Directional lighting may help to reveal details in the task.

Illuminance values may be varied to suit circumstances; see section 2.3.2, Illuminance.

Public car parks (indoor)

	Maintained illuminance (lux)	Limiting glare rating	Minimum colour rendering (R_a)	Notes
In/out ramps (during the day)	300	25	20	1, 2
In/out ramps (at night)	75	25	20	1, 2
Traffic lanes	75	25	20	1, 2
Parking areas	75	–	20	1, 2, 3
Ticket office	300	19	80	4, 5

Notes:

1. The illuminance should be provided at floor level.
2. Safety colours should be recognisable.
3. A high vertical illuminance increases recognition of people's faces and therefore the feeling of safety.
4. Avoid reflections in the windows.
5. Prevent glare from outside.

Illuminance values may be varied to suit circumstances; see section 2.3.2, Illuminance.

Railway installations

	Maintained illuminance (lux)	Limiting glare rating	Minimum colour rendering (R_a)	Note
Covered platforms and passenger subways (underpasses)	50	28	40	1
Open platforms	10	–	–	1
Ticket hall and concourse	200	28	40	
Ticket and luggage offices and counters	300	19	80	
Waiting rooms	200	22	80	

Note:

1. Care should be taken to light and clearly to mark the platform edge.

Illuminance values may be varied to suit circumstances; see section 2.3.2, Illuminance.

Rest, sanitation and first aid rooms

	Maintained illuminance (lux)	Limiting glare rating	Minimum colour rendering (R_a)	Notes
Canteens, pantries	200	22	80	1, 2
Rest rooms	100	22	80	3
Rooms for physical exercise	300	22	80	
Cloakrooms, washrooms, bathrooms, toilets	200	25	80	4
Sick bay	500	19	80	
Rooms for medical attention	500	16	90	5

Notes:

1. The lighting should aim to provide a relaxed but interesting atmosphere.
2. In food storage areas, luminaires should be capable of being washed or hosed down in safety.

3. Lighting should be different in style from that in the work areas.
4. In bathrooms, luminaires must be suitable for damp and humid situations.
5. Colour temperature to be at least 4000 K.

Illuminance values may be varied to suit circumstances; see section 2.3.2, Illuminance.

Retail premises

	Maintained illuminance (lux)	Limiting glare rating	Minimum colour rendering (R_a)	Note
Sales area	300	22	80	1
Till area	500	19	80	
Wrapper table	500	19	80	

Note:

1. Both illuminance and UGR requirements are determined by the type of shop.

Illuminance values may be varied to suit circumstances; see section 2.3.2, Illuminance.

Rolling mills, iron and steel works

	Maintained illuminance (lux)	Limiting glare rating	Minimum colour rendering (R_a)	Notes
Production plants without manual operation	50	–	20	1, 2
Production plants with occasional manual operation	150	28	40	2
Production plants with continuous manual operation	200	25	80	3
Slab store	50	–	20	1
Furnaces	200	25	20	1
Mill train; coiler; shear line	300	25	40	
Control platforms: control panels	300	22	80	
Test, measurement and inspection	500	22	80	
Under-floor man-sized tunnels; belt sections; cellars etc.	50	–	20	1, 2

Notes:

General: lamps and luminaires may be subject to severe fouling and high ambient temperatures. Dustproof (IP5X) or other luminaires with good maintenance properties are desirable in areas where raw materials are formed into the basic product.

1. Safety colours should be recognisable.
2. Supplementary lighting may be required for maintenance work.
3. If high-bay lighting is used the colour rendering requirement may be relaxed provided that measures are taken to ensure lighting with higher colour rendering is provided at continually occupied work stations.

Illuminance values may be varied to suit circumstances; see section 2.3.2, Illuminance.

Store rooms, cold stores

	Maintained illuminance (lux)	Limiting glare rating	Minimum colour rendering (R_a)	Notes
Store and stockrooms	100	25	60	1, 2
Dispatch packing handling areas	300	25	60	
Automatic high rack stores – unmanned gangways	20	–	40	3, 4
Automatic high rack stores – manned gangways	150	22	60	4
Automatic high rack stores – control stations	150	22	60	5

Notes:

1. 200 lux if continuously occupied.
2. If small items that are visually difficult to identify are stored, then 300 lux and supplementary local lighting may be needed.
3. Supplementary lighting may be required for maintenance.
4. Illuminance at floor level.
5. Avoid glare to operator, local lighting should be considered.

Note that in cold stores the light source should be chosen to be adequate for reliable starting at −30°C. Illuminance values may be varied to suit circumstances; see section 2.3.2, Illuminance.

Textile manufacture and processing

	Maintained illuminance (lux)	Limiting glare rating	Minimum colour rendering (R_a)	Notes
Work places and zones in baths, bale opening	200	25	60	
Carding, washing, ironing, devilling, machine work, drawing, combing, sizing, card cutting, pre-spinning, jute and hemp spinning	300	22	80	
Spinning, plying, reeling, winding	500	22	80	1
Warping, weaving, braiding, knitting	500	22	80	1
Sewing, fine knitting, taking up stitches	750	22	80	
Manual design, drawing patterns	750	22	90	2
Finishing, dyeing	500	22	80	
Drying room	100	28	60	
Automatic fabric printing	500	25	80	
Burling, picking, trimming	1000	19	80	
Colour inspection; fabric control	1000	16	90	2
Invisible mending	1500	19	90	2
Hat manufacture	500	22	80	

Notes:

1. Care should be taken to minimise stroboscopic effects on rotating machinery.
2. Colour temperature of the light should be greater than 4000 K.

Illuminance values may be varied to suit circumstances; see section 2.3.2, Illuminance.

Theatres, concert halls and cinemas

	Maintained illuminance (lux)	Limiting glare rating	Minimum colour rendering (R_a)	Notes
Practice rooms, dressing rooms	300	22	80	1
Foyers	200	–	–	
Booking offices	300	22	80	2
Auditoria	100	–	–	3
Projection rooms	150	22	40	4

Notes:

1. Lighting of mirrors for make-up shall be glare-free.
2. Local or localised lighting may be appropriate.
3. Dimming facilities will be necessary.
4. Lighting should be provided on the working side of the projector. The lighting should not detract from the view into the auditorium. Dimming facilities may be desirable.

Illuminance values may be varied to suit circumstances; see section 2.3.2, Illuminance.

Trade fairs and exhibition halls

	Maintained illuminance (lux)	Limiting glare rating	Minimum colour rendering (R_a)
General lighting	300	22	80

Illuminance values may be varied to suit circumstances; see section 2.3.2, Illuminance.

Vehicle construction

	Maintained illuminance (lux)	Limiting glare rating	Minimum colour rendering (R_a)	Note
Body work and assembly	500	22	80	
Painting, spraying chamber, polishing chamber	750	22	80	
Painting: touch up, inspection	1000	19	90	1
Upholstery manufacture (manned)	1000	19	80	
Final inspection	1000	19	80	

Note:

1. Colour temperature of the light should be greater than 4000 K.

Illuminance values may be varied to suit circumstances; see section 2.3.2, Illuminance.

Wood working and processing

	Maintained illuminance (lux)	Limiting glare rating	Minimum colour rendering (R_a)	Notes
Automatic processing (e.g. drying, plywood manufacture)	50	28	40	1
Steam pits	150	28	60	2
Saw frame	300	25	80	
Work at joiner's bench, gluing, assembly	300	25	80	
Polishing, painting, fancy joinery	750	22	80	1
Work on wood-working machines (e.g. turning, fluting, dressing, rebating, grooving, cutting, sawing, sinking)	500	19	80	1, 2
Selection of veneer woods	750	22	90	3
Marquetry, inlay work	750	22	90	3
Quality control, inspection	1000	19	90	3

Notes:

1. Dust from sanding and similar operations may represent an explosion hazard; luminaires should be chosen appropriately.
2. Care should be taken to minimise stroboscopic effects on rotating machinery.
3. Colour temperature of the light should be greater than 4000 K.

Illuminance values may be varied to suit circumstances; see section 2.3.2, Illuminance.

Part 3 Lighting design

The flow diagram shown in Figure 3.1 presents a design approach, based on reasonable practice, to applying the principles, recommendations and technology described elsewhere in the *Code*. With experience, the lighting designer will develop this to arrive at individual design solutions. This reference to 'the lighting designer' should not be read to imply that the design process can be carried out in isolation. All successful lighting is the result of close collaboration and interaction within the building design team, to interpret the client's brief. The purpose of this section, however, is to concentrate on the lighting designer's responsibility and contribution to the total design process.

The design process is described in detail in the following sections.

3.1 Objectives

The first stage in planning is to establish the lighting design objectives which guide the decisions in all the other stages of the design process. It is a matter of deciding for what, and for whom, the lighting is intended, rather than referring at this stage to the light-

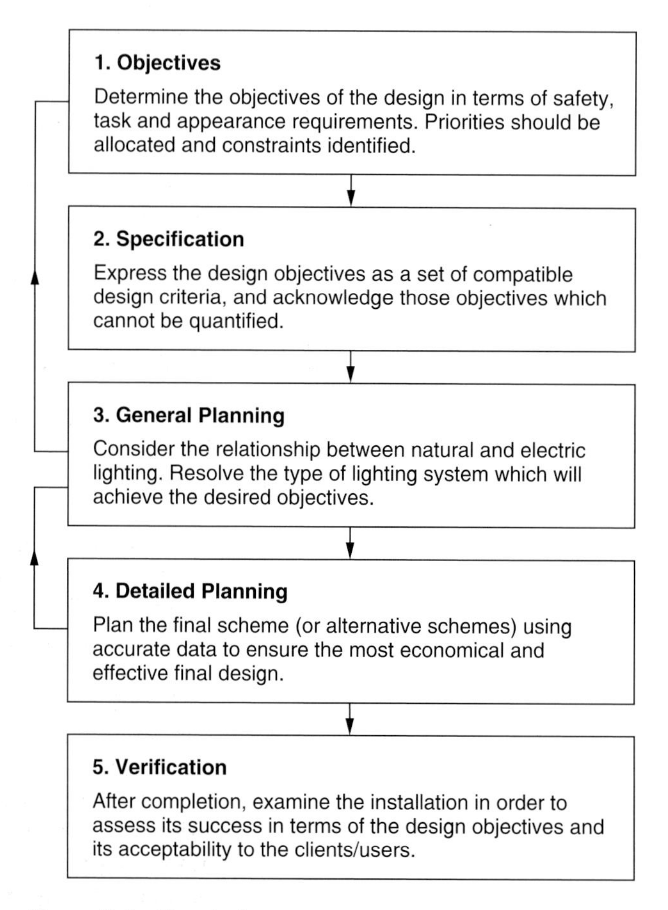

Figure 3.1 The design process

ing schedule. These objectives can be considered in three parts. The objectives must be balanced and priorities set bearing in mind the constraints that apply to the given task.

3.1.1 Safety

The lighting installation must be electrically and mechanically safe and must allow the occupants to use the space safely. These are not only primary objectives but also statutory obligations. It is, therefore, necessary to identify any hazards present and the need for emergency lighting (see section 3.8.6, Emergency lighting).

3.1.2 Visual tasks

The type of work which takes place in the interior will define the nature and variety of the visual tasks. An analysis of the visual tasks (there is rarely just one) in terms of size, contrast, duration and need for colour discrimination is essential to establish the quantity and quality of the lighting required to achieve satisfactory visual conditions.

In a 'general' office, for example, at one extreme the task may be to answer the telephone (a simple visual task). At the other extreme the occupants may have to transcribe text, handwritten in pencil, using VDTs. This presents a complex set of visual tasks. In addition to establishing the nature of the tasks it is also necessary to identify the positions and planes on which the tasks lie. This information is essential if lighting matched to the tasks is to be provided.

3.1.3 Appearance and character

It is necessary to establish what mood or atmosphere is to be created. This is not just for prestige offices, places of entertainment, and the like, but should be considered in all designs, even where it will be given less importance than other factors.

3.1.4 Priorities and constraints

The above objectives will not have equal weight. Some will be 'essential' while others can only be 'desirable'. The evaluation will depend upon the priorities and constraints set by the client or the application.

Often the most obvious constraint is financial. Few clients want to spend more than is necessary to meet their brief and objectives. Different clients will allocate more or less to meet the various objectives according to their own valuation of the final result. This will temper and modify the importance of the various design objectives, but should be opposed if solely financial consideration suppresses any of the essential requirements of the design solutions.

Both capital expenditure and running costs should be considered to achieve the most economical scheme. This does not always happen because a second system of budget control applies to the running cost. This is an unsatisfactory approach and should, if possible, be resisted. Capital and running costs should be taken together to establish the lowest overall investment (see Financial appraisals on CD).

Other constraints which may affect the design objectives are:

— energy consumption
— hazardous or onerous environmental considerations (which may limit the range of acceptable luminaires)
— physical problems affecting the installation of equipment
— access for maintenance.

These constraints must be recognised when setting the objectives of the design.

3.2 Specification

The designer must always take due note of statutory instruments that affect lighting conditions. Lighting designers have a responsibility to ensure that lighting is not liable to cause injury to the health of occupants. Bad lighting can contribute towards accidents or result in inadequate working conditions.

The lighting objectives now need to be expressed in a suitable form. Although many can be expressed in physical terms, suitable design techniques may not exist or may be too cumbersome. For example, obstruction losses (see section 3.8.4, Specification and interpretation of illuminance variation) and contrast-rendering factors are two quantities that are difficult to calculate and predict accurately. Not all design objectives can be expressed as measurable quantities. For example, the need to make an environment appear 'prestigious', 'efficient' or 'vibrant' cannot be quantified. This does not mean that these objectives should be ignored, but experience and judgement may have to replace calculation.

A full specification can be established by reference to Part 2 of this *Code* and by taking the design objectives into account.

3.3 General planning

The remaining stage of design is to translate the design specification into the best possible solution, to meet the original objectives. The specification is therefore only a stepping stone; if it proves difficult to plan an installation that meets the design specification, it may be necessary to reassess the original objectives.

At the general planning stage, as distinct from detailed planning (see section 3.8, Detailed planning), the designer aims to establish whether the original objectives are viable, and to resolve what type of design can be employed to satisfy these objectives. The initial stage in the general planning of a lighting installation is to consider the interior to be lit, its proportions, its contents, and the daylight available. The first two topics covered here are:

— daylight
— choice of electric lighting systems.

3.4 Daylight

The use of daylighting and electric lighting together can both contribute towards the efficient use of primary energy and increase the satisfaction of users. The specification of daylight requirements is covered in *BS 8206* Part 2, and the design of windows is described in the *CIBSE Lighting Guide 10: Daylighting and Window Design.*

The energy saving from daylight use can be estimated from the average daylight factor. The analysis of energy use is described in section 2.4, Energy efficiency recommendations, while information on control systems for electric lighting used in conjunction with daylight is provided in Lighting controls (see CD) and section 3.7, Energy management.

When working on an existing building, or when the window design of a new building has already been fixed, the designer should take the following steps:

(*a*) Check whether any form of shading devices or blinds will be required to control sunlight penetration.

(*b*) Analyse the extent to which daylight will provide general room lighting or task lighting.

(*c*) Design the electric light for daylight hours and for night.

(*d*) Select control equipment to ensure efficient use of electric lighting.

3.4.1 Initial appraisal of daylight quantity

The extent of daylight penetration can be assessed initially using the following guidelines:

(*a*) If the sky is not directly visible from a point in an interior, the level of daylight at that point will be small. The 'no-sky line' (the boundary of the region in the room from which no sky can be seen) gives an indication of the area beyond which daylight may not contribute to general room lighting. Windows can, however, still be important in providing an external view from parts of a room distant from the window walls.

(*b*) In areas of a room adjacent to a side window, the region in which daylight might contribute significantly towards task lighting extends back from the window for a distance of about twice the height of the window head above the working plane (provided that there are not large external obstructions, that there is clear glazing, and that the sill is not significantly higher than the working plane).

3.4.2 Daylight to enhance the general brightness of the room

The use of daylight for general room lighting is described in sections 1.2, Daylight and electric light, and 2.2, Recommendations for daylighting. The extent to which existing windows may give a daylit appearance and also provide the necessary overall brightness of the room can be found by calculating the average daylight factor. This value can then be compared with the criteria given in section 2.2.1, Daylight for general room lighting.

The average daylight factor may be calculated as follows:

$$D = \frac{TA_w\theta}{A(1 - R^2)}$$

where T is the diffuse light transmittance of the glazing, including the effects of dirt, blinds, curtains and any other obstructions or coverings; A_w is the net glazed area (in m^2) of the window; and θ is the angle (in degrees) subtended by the visible sky (θ is measured in a vertical plane normal to the glass, from the window reference point, as illustrated in Figures 3.2 and 3.3); A is the total area (in m^2) of the interior surfaces – ceiling, walls, windows, floor; R is the area-weighted average reflectance of these interior surfaces (in initial calculations for rooms with white ceilings and mid-reflectance walls, this may be taken as 0.5).

The equation should not be applied where external obstructions cannot be represented by a single angle of elevation, such as where a window faces into a courtyard. Alternative calculation methods are available for complex cases.

If all the windows in a room have the same transmittance and face the same angle of obstruction, the average daylight factor may be found at once by letting A_w be the total glazed area. Otherwise the average daylight factor should be found for each window separately, and the results summed.

3.4.3 Daylight for task illumination

The daylight illuminance (E_{in}) at a point in a room may be estimated using the following equation:

$$E_{in} = E_h f_o D$$

where E_h is the external unobstructed horizontal illuminance in lux (see Figures 3.4 and 3.5); f_o is a window orientation factor (this allows for the effects of window orientation – see Table 3.1); D is the daylight factor at the point in the room, expressed as a fraction (i.e. the percentage value divided by 100).

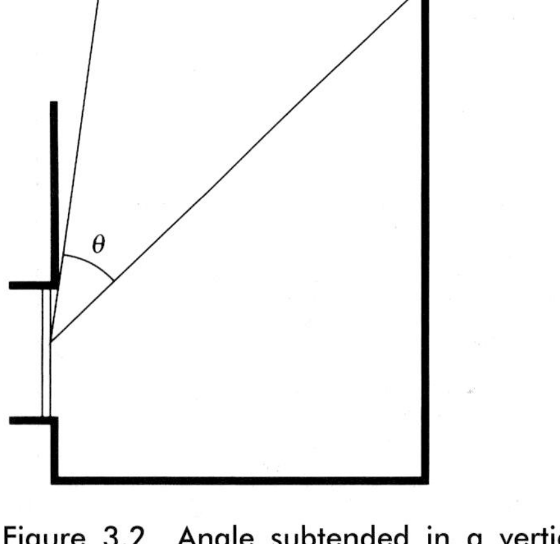

Figure 3.2 Angle subtended in a vertical plane normal to the sky; θ is the angle subtended in a vertical plane normal to the window, by sky visible from the centre of the window

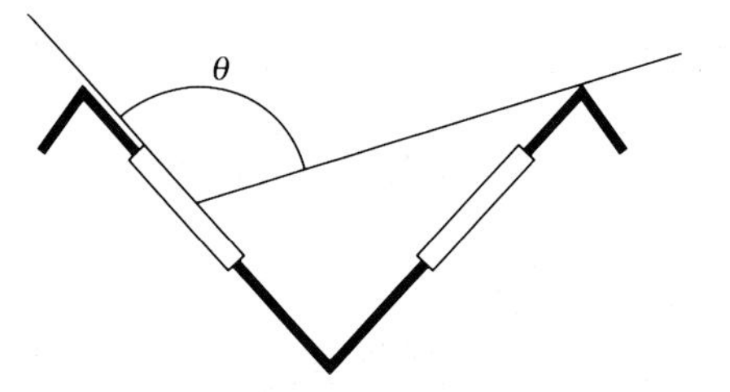

Figure 3.3 Angle subtended in a vertical plane normal to the rooflight; θ is the angle subtended in a vertical plane normal to the rooflight, by sky visible from the centre of the rooflight

Table 3.1 Diffuse orientation factors for a working day, 0900 hours to 1700 hours

Orientation	Orientation factor (f_o)
North	0.97
East	1.15
South	1.55
West	1.21
Horizontal rooflight	1.00

Diffused illuminance (E_h, klx) availability for Edinburgh.

Methods for calculating the point daylight factor are given in *BRE Digest 303: Estimating Daylight in Buildings*, and in the CIBSE publication *Lighting Guide 10: Daylighting and Window Design*.

The factors for other orientations may be obtained by interpolation. The factors given may be applied with reasonable accuracy for working days finishing between 1600 hours and 1900 hours (see Figures 3.4 and 3.5).

Figures 3.4 and 3.5 give the availability of daylight in London and Edinburgh for various lengths of the working day. The graph for London should be applied to sites in southern and central England; the Edinburgh graph should be applied to sites in Scotland, northern England and Northern Ireland.

For example:

An illuminance of 500 lux is required on a desk for a working day of 0900–1700 hours. The daylight factor at the desk, from an east-facing window, is 0.015, or 1.5 per cent. The building is in southeast England.

The orientation factor is 1.15. The external illuminance required to give 500 lux on the desk is $500/(1.15 \times 0.015) =$ 28 986 lux, or approximately 29 klux. From the London graph (Figure 3.4) it can be seen that this level is achieved for 28 per cent of working hours.

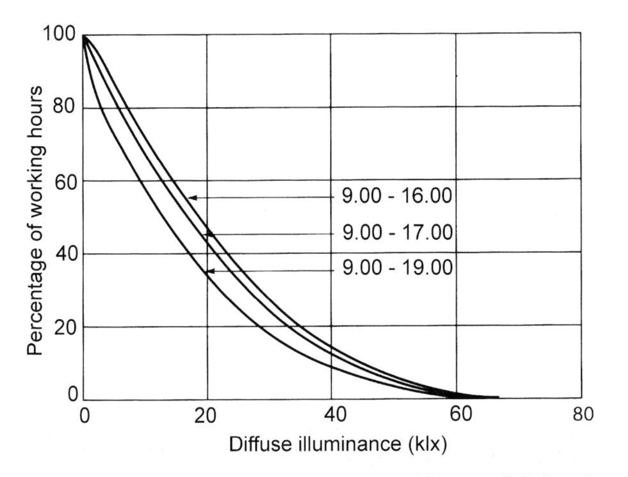

Figure 3.4 Diffuse illuminance (E_h, klx) availability for London

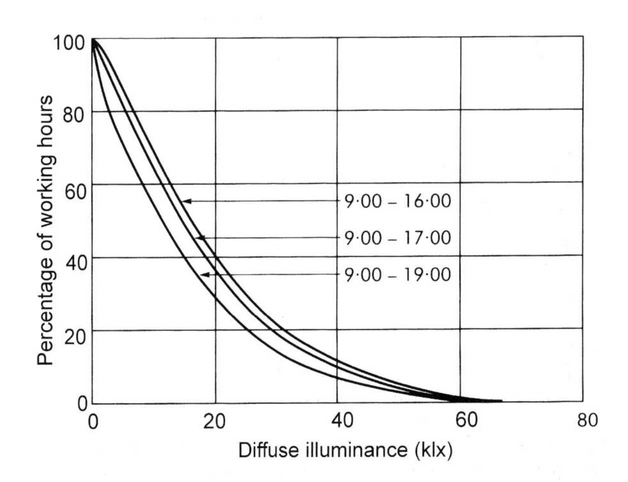

Figure 3.5 Diffuse illuminance (E_h, klx) availability for Edinburgh

3.5 Choice of electric lighting systems

3.5.1 General lighting

Lighting systems that provide an approximately uniform illuminance over the whole working plane are called general lighting systems (see Figure 3.6). The luminaires are normally arranged in a regular layout, and the appearance of the installation is usually tidy but may be rather bland. General lighting is simple to plan using the lumen method (see section 3.5.3, Average illuminance (lumen method)) and requires no coordination with task locations. The greatest advantage of such systems is that they permit flexibility of task location.

If the installation design assumes an empty space between ceiling and working plane, it may be difficult to achieve the recommended uniformity if the area actually contains substantial obstruction or is divided into several small areas. If the degree of obstruction over the working plane is high, or if partitioning is installed, it will probably be necessary to increase the number of luminaires. This will have significant energy and economic implications, especially if a larger number of lower wattage luminaires are required.

The major disadvantage of general lighting systems is that energy may be wasted in illuminating the whole area to the level needed for the most critical tasks. Energy could be saved by providing the necessary illuminance over only the task areas, and using a lower ambient level for circulation and other non-critical tasks.

(a)

(b)

Figure 3.6(a) and (b) A general lighting system employs a regular array of luminaires to provide a uniform illuminance across the working plane

3.5.2 Localised lighting

Localised lighting systems (see Figure 3.7) employ an arrangement of luminaires designed to provide the required maintained illuminance on work areas together with a lower illuminance for the other areas. The average illuminance on the other areas should not be less than one-third of the average illuminance over the work areas (see sections 1.4, Variation in lighting, and 2.3.4, Luminance and illuminance ratios).

The lighting layout must be coordinated with the task positions and orientation. The system can be inflexible, and information on plant and furniture layout is essential at the design stage. Changes in the work layout can seriously impair a localised system, although uplighters and other easily relocatable systems or energy management controls (see section 3.7, Energy management) can overcome these problems.

Localised systems normally consume less energy than general lighting systems unless a high proportion of the area is occupied by workstations. This should be confirmed by calculations. Maintenance of localised systems can be more critical than with general lighting systems.

(a)

(b)

Figure 3.7(a) and (b) A localised lighting system uses luminaires located adjacent to the workstations to provide the required task illuminance. The necessary ambient illuminance in the surrounding areas is provided by additional luminaires if required

3.5.3 Local lighting

Local lighting provides illumination only over the small area occupied by the task and its immediate surroundings (see Figure 3.8). A general lighting system must be installed to provide sufficient ambient illumination for circulation and non-critical tasks. This is then supplemented by the local lighting system to achieve the necessary design maintained illuminance over the task areas. The general surround average illuminance should not be less than one-third of the average task illuminance (see sections 1.4, Variation in lighting, and 2.3.4, Luminance and illuminance ratios). Local lighting can be a very efficient method of providing adequate task illumination, particularly where high illuminances are necessary or flexible directional lighting is required. Local lighting is frequently provided by luminaires mounted at the work place in offices and factories.

Local lighting must be positioned to minimise shadows, veiling reflections and glare. Although local luminaires allow efficient utilisation of emitted light, the lower wattage lamp circuits will be less efficient and the luminaires can be expensive. Most local lighting systems are accessible and often adjustable. This increases wear and tear and hence maintenance costs, but the system provides individual control, which is often favoured by those working in the area.

Both local and localised lighting offer scope for switch control of individual luminaires that can be off when not required, but sufficient ambient illumination must be provided at all relevant times.

(a)

(b)

Figure 3.8(a) and (b) A local lighting system employs a general lighting scheme to provide the ambient illuminance for the main area, with additional luminaires located at the workstations to provide the necessary task illuminance

3.6 Choice of lamp and luminaire

The choice of lamp will affect the range of luminaires available, and vice versa. Therefore, one cannot be considered without reference to the other. One method of design is to follow a procedure that does not start with identifying a single lamp and luminaire combination, but rather rejects those combinations that are unsatisfactory. In this manner, whatever remains will be acceptable and a final choice can be made by comparison. With such an approach, if all available choices are eliminated this probably indicates that one or more of the objectives are unrealistic. Finally, all the un-rejected luminaire and lamp combinations are acceptable, and the most efficient, economical and architecturally acceptable scheme can then be selected.

3.6.1 Selection of lamp characteristics

The designer should compile a list of suitable lamps by rejecting those that do not satisfy the design objectives. For general guidance, see Lamps (see CD). However, up-to-date manufacturers' data should be used for final selections.

The run-up times of all but low-pressure fluorescent discharge lamps are unsatisfactory for applications requiring instant illumination when switched on, unless auxiliary tungsten or fluorescent lamps are provided.

Lamps must have colour-rendering properties suited to their intended use. Good colour rendering may be required in order to achieve better discrimination between colours where this is part of the visual task. Alternatively, good colour rendering may be required to achieve a particular appearance or degree of comfort (e.g. in offices, merchandising, or leisure activities). The choice of colour appearance can be used by the designer to create an appropriate 'atmosphere'. For example, 'warm' colour appearance might be selected for informal situations, whereas a 'cold' appearance could be associated with formality. This is an entirely subjective judgement, but adjacent areas should not be lit with sources of significantly different apparent colour unless a special effect is required.

The life and lumen maintenance characteristics of the lamps must be considered to arrive at a practicable and economic maintenance schedule.

Where moving machinery is used, care should be taken to avoid stroboscopic effects. All lamps operating on an alternating current exhibit some degree of cyclic variation of light output. It is most significant with discharge lamps that do not employ a phosphor coating. The problem can normally be reduced or eliminated by having alternate rows of luminaires on different phases of the supply and ensuring that critical areas receive illumination in roughly equal proportion from each phase. Alternatively, some lamps may be operated from high frequency electronic ballasts, or illumination from local luminaires (with acceptable lamps that do not cause stroboscopic problems) can be used to swamp the general illumination.

One other factor that may be a limitation on the use of certain lamp or circuit types is minimum starting temperature. Particularly in the case of linear fluorescent lamps, this is also influenced by the luminaire design.

When selecting a range of suitable lamps, the designer must consider the types of luminaires that are available and the degree of light control and light output required. Accurate light control is more difficult with large area sources than with small area sources; however, the latter will have a higher luminance (for the same output) and are potentially more glaring.

Standardisation of a limited selection of lamp types and sizes without compromising visual requirements for a particular site or company can simplify maintenance and stocking.

3.6.2 Selection of luminaire characteristics

In addition to being safe, luminaires may have to withstand a variety of physical conditions – e.g. vibration, moisture, dust, ambient temperature, or vandalism. Also, the external appearance of the luminaire, its fixing and location must be in sympathy with the architectural style of the interior. General guidance on the characteristics of luminaires can be obtained from Luminaires, see Sony CD.

Safety is assured by using equipment meeting the required standards, as in Luminaire standards and marking, and in Luminaire quality systems and approval marks (see CD).

Luminaire reliability and life will have a direct impact on the economics of the scheme. The ease with which luminaires can be installed and maintained (Maintenance of lighting installations, see CD) will also affect the overall economics and convenience of the scheme. For example, luminaires that can be unplugged and detached, or that have removable gear, can simplify maintenance by allowing remote servicing.

Not only must the luminaire withstand the ambient conditions, it may also have to operate in a hazardous area such as a refinery, mine or similar environment. In this event, special equipment is required to satisfy the safety regulations. This subject is covered by the CIBSE *Lighting Guide: Lighting in Hostile and Hazardous Environments*.

The light distribution of the luminaire influences the distribution of luminance and the directional effects that will be achieved. The illuminance ratios are described in Luminaire characteristics (see CD) and section 3.6.3, Illuminance ratio charts, for a regular array of a given luminaire, and can be calculated by a number of methods. Currently the method given in CIBSE *Technical Memorandum 5* is recommended, although it is likely that *Technical Memorandum 5* will be replaced shortly.

The utilisation factor (UF) for a luminaire is a measure of the efficiency with which light from the lamp is used for illuminating the working plane (see section 3.8.3, Average illuminance (lumen method).

For a given interior and set of environmental conditions, the lamp, circuit and luminaire performance will influence the installed power of the lighting system. The power density (W/m^2) of alternative solutions should be compared with the target ranges given in section 2.4.3, Lighting energy targets. Nevertheless, the system of the lowest installed load will not necessarily achieve the lowest energy use if a greater degree of energy management control can be achieved with one type of lamp rather than another (see section 3.7, Energy management).

In addition to the practical performance of the luminaire, consideration must be given to its appearance. This can range from the totally discreet, such as a fully recessed, low brightness downlighter, to a highly decorative chandelier. The choice of luminaire style and degree of design expression will be strongly influenced by architectural and interior design considerations. Skill and care are required in the selection or specification of luminaires that satisfy aesthetic criteria whilst performing efficiently and safely.

3.6.3 Illuminance ratio charts

Illuminance ratio (IR) charts were published in CIBSE (IES) *TR15*. They enable the designer to examine the effects of room index, surface reflectances, luminaire direct ratio (DR) and flux fraction ratio (FFR) upon illuminance ratios and the directional aspects of lighting.

They were presented in pairs for different combinations of ceiling, wall and floor reflectance and for different room indices. The reflectances are given the symbols L, M and D, to signify light, medium and dark reflectances, respectively (see Table 3.2).

Table 3.2 Reflectances for room surfaces

	L	M	D
Ceiling cavity	0.70	0.50	0.30
Walls	0.50	0.30	0.10
Floor cavity	0.30	0.20	0.10

Figure 3.9 shows a typical pair of charts. The charts are identical for each room index except for the loci plotted on them. In each case the horizontal axis represents the DR of the installation and the vertical axis is the FFR fraction ratio (FFR) of the installation. Luminaires can therefore be plotted onto the charts according to their FFR and DR. The DR is calculated from the distribution factor, for the working plane or floor (DF_F), of the luminaire for an appropriate room index (RI) and the downward light output ratio (DLOR) of the luminaire.

$$DR = \frac{DF_F}{DLOR}$$

The value of DF_F is the same as the utilisation factor (UF) of the chosen luminaire at zero reflectance. Loci of constant illuminance ratio are plotted on the left-hand chart and loci of constant average vector/scalar ratio are plotted on the right-hand chart (Illuminance vector, and Scalar illuminance – see CD).

The charts may be used in two ways. Luminaires can be plotted onto the charts to determine the illuminance ratios and average vector/scalar ratios that will be achieved by a regular array of such luminaires. Alternatively, at the general planning stage, the charts may be used to identify the range of reflectances and luminaires that can achieve the desired conditions. The range of acceptable luminaires can be identified by DR and FFR. The process of selection will be aided if the

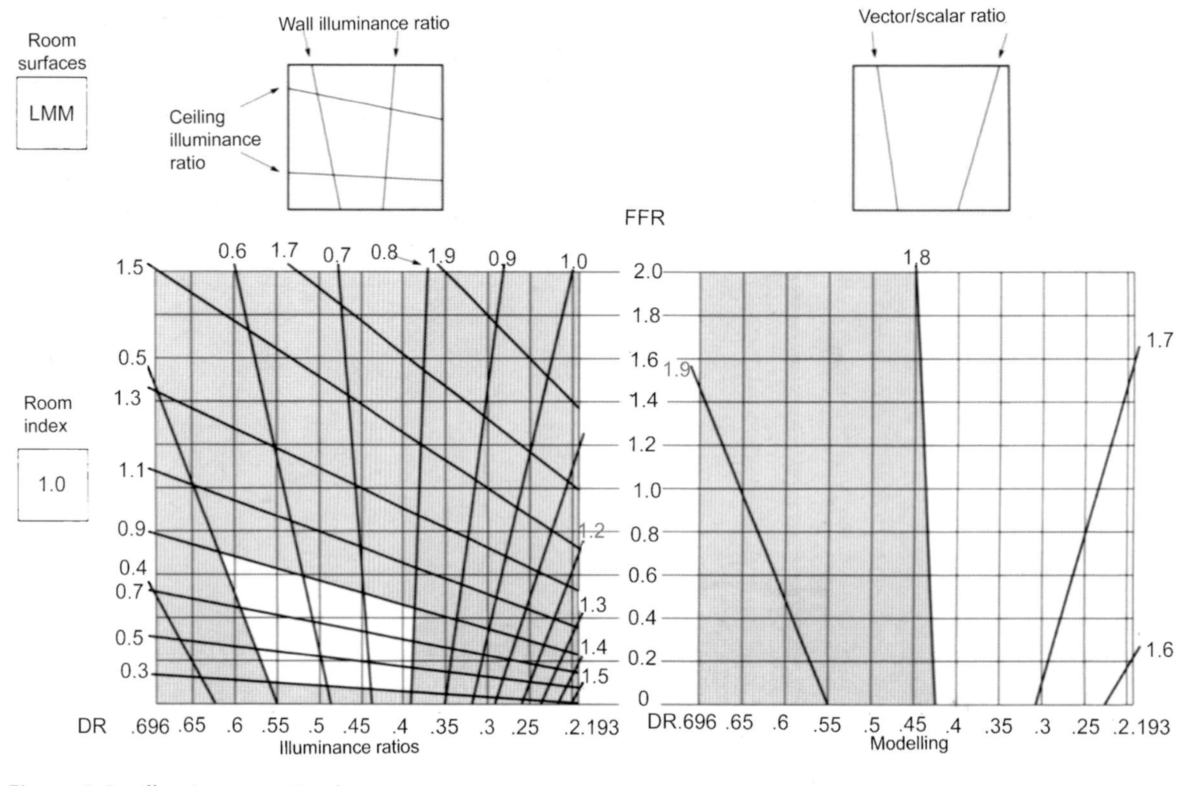

Figure 3.9 Illuminance ratio charts

positions of luminaires are marked onto the charts or a transparent overlay.

Acceptable ranges of illuminance ratios and vector/scalar ratios are shown as unshaded (safe) areas on the charts. These values are not sacrosanct, and there are reasons why a designer may wish to deviate from them. Bright walls can make a room seem larger and more spacious, whereas dark walls can make it seem small and possibly cramped, or intimate. Bright ceilings and dark walls may give the impression of formality and tension, whilst the reverse (bright walls and dark ceiling) may create an informal and relaxed or sociable atmosphere. These are not hard and fast rules, but are supported by experimentation. Figure 3.10 shows these tendencies mapped onto a typical IR chart.

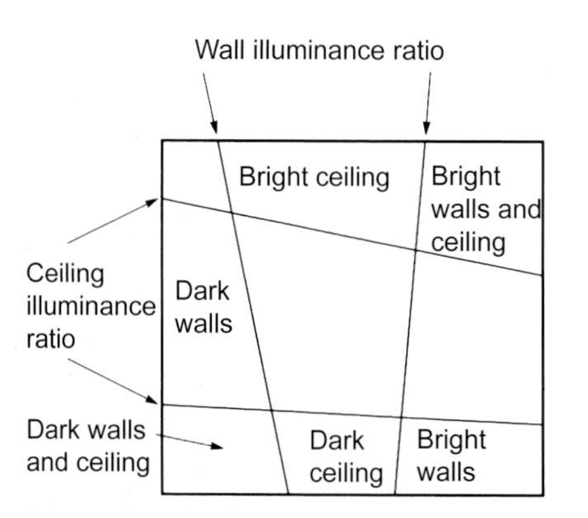

Figure 3.10 The effect of illuminance ratios on room appearance

It is often impossible simultaneously to achieve the desired illuminance ratios and vector/scalar ratios without changes in reflectance, and even then it may still be impossible. Wall-washers may be employed to increase the wall-to-task illuminance ratio. The IR charts illustrate that for most cases a proportion of upward light from the luminaire is desirable to achieve acceptable ceiling-to-task illuminance ratios. Table 3.3 shows a typical table of utilisation factors.

Table 3.3 Typical presentation of utilisation factor table and associated performance data

Nadir intensity*	302 cd/1000 lm
SHR_{max} (square)	1.36
$SHR_{max\ tr}$ (continuous rows)	1.75
ULOR*	0.00
DLOR*	0.64
LOR*	0.64

T A

***Correction factors**

	36 W	58 W	32 W	50 W
Length Factor	1.00	1.00	1.00	1.00
HF Factor	—	—	1.01	1.01

Utilisation Factors (UF_F) for SHR_{nom} of 1.25

Room Reflectances			Room index (K)								
C	**W**	**F**	**0.75**	**1.00**	**1.25**	**1.50**	**2.00**	**2.50**	**3.00**	**4.00**	**5.00**
70	50	20	0.45	0.51	0.56	0.58	0.62	0.64	0.66	0.68	0.69
	30		0.41	0.48	0.52	0.55	0.59	0.62	0.64	0.66	0.68
	10		0.38	0.45	0.49	0.53	0.57	0.60	0.62	0.65	0.66
50	50	20	0.44	0.50	0.54	0.57	0.60	0.62	0.64	0.65	0.67
	30		0.40	0.47	0.51	0.54	0.58	0.60	0.62	0.64	0.65
	10		0.38	0.44	0.49	0.52	0.56	0.58	0.60	0.63	0.64
30	50	20	0.43	0.49	0.53	0.55	0.58	0.60	0.62	0.63	0.64
	30		0.40	0.46	0.50	0.53	0.56	0.59	0.60	0.62	0.63
	10		0.37	0.44	0.48	0.51	0.55	0.57	0.59	0.61	0.62
0	0	0	0.36	0.43	0.47	0.49	0.53	0.55	0.56	0.58	0.59

Using the photometric data given above: DLOR = 0.64, DF_F (room index K = 1.0) = 0.43, therefore DR = 0.67. As the ULOR = 0%, the FFR = 0.

By plotting these values, which apply to reflectances LMM and K = 1.0, both the illuminance ratios and vector/scalar ratio are outside the 'safe area'. It can be seen that an installation of this luminaire in a small room would produce dark walls and ceilings.

Note: The charts in TR15 are based on the British Zonal method (BZ) numbers and charts have been re-plotted for direct ratios. The figures below can be used to produce overlays for use with the IR charts in the last edition of TR15

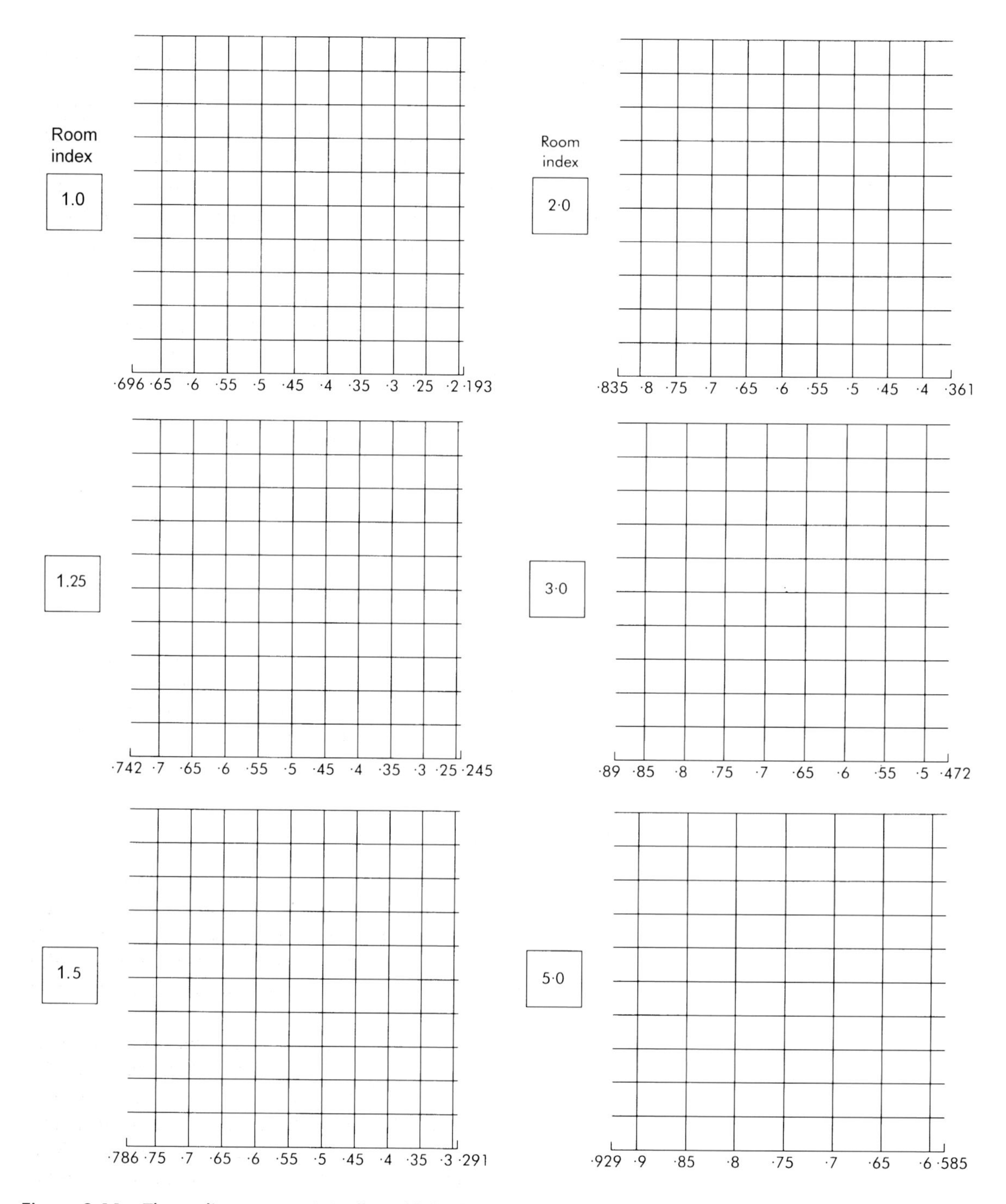

Figure 3.11 These diagrams, originally published in IES Technical Report 15, can be used to produce overlays for use with the IR charts in that document.

3.7 Energy management

A lighting system must be designed and managed to achieve good control of energy use. This is important during the working day and outside working hours.

3.7.1 Choice of controls

The factors that influence the specification of controls include occupancy, occupancy pattern, available daylight, type of lighting (i.e. can it be dimmed?), the desired level of control sophistication and, of course, costs.

The cost of a control system installation should be compared with the cost of a traditional hard-wire installation, and the difference related to the projected energy savings. Especially with new buildings, the cost difference may be very small. For existing installations there may be constraints on selection of controls where the existing wiring gives little scope for alteration or change. The use of mains-borne signalling may reduce these constraints and allow a central system to be installed without disturbing existing wiring, but it is essential to ensure compatibility with other electrical and signalling circuits. Simple reset switches may also be installed without significantly affecting existing wiring.

Alternatively, the use of self-contained luminaires, each with its own sensor, may be a more practical and economic solution than centralised control. For this it has to be accepted that certain refinements of centralised control cannot be achieved.

The following control elements can be considered.

3.7.1.1 Daylight linking

One or more of the lighting rows adjacent to the windows (see section 3.4.1, Initial appraisal of daylight quantity) may be linked to either external or internal photocells to monitor daylight and adjust the electric lighting accordingly, by either switching or dimming.

3.7.1.2 Constant illuminance

Designing for maintained illuminance means that initially, when lamps are new and luminaires and room surfaces are clean, the illuminance will be substantially higher than the design level. How much higher will depend on the characteristics of the installation and the maintenance programme that the user intends to follow.

High frequency fluorescent lamp systems, which can be regulated, can be linked to photocells that will hold the lighting at the design maintained illuminance value. As the system ages, the controls will automatically increase the power to the lamp. Eventually the system will operate at full load in order to produce the maintained illuminance, and this is when maintenance should be carried out.

The same control system can also cover change of use. If the function of an area changes, requiring a lower task illuminance, the system can be adjusted to control the lighting to the revised level.

3.7.1.3 Occupancy

Lighting linked to occupancy, or more appropriately to occupancy pattern, can show considerable savings in energy usage.

An example of occupancy detection is where a detector senses the approach of a forklift truck and switches lighting between warehouse aisles. A predetermined time delay should be built into the control system to avoid excessive switching, which can shorten lamp life.

This form of control can be applied to a wider range of lamp types, as long as the run-up and re-strike characteristics (Lamps, see LIF Lamp guide on CD) are taken into account.

3.7.1.4 Automatic control

Automatic control can take a number of forms.

A timer control system may switch the whole lighting installation on and off at predetermined times, or it may be programmed to send signals at certain times during the day (e.g. lunch time) to switch off selected luminaires. If daylight is sufficient, or lights go out over unoccupied areas, it is unlikely that these will be switched on again until needed. This type of system can be used for providing reduced lighting levels early in the morning and before the majority of staff arrive, or in the evening to cover cleaning or security operations. Local manual override switching is essential with this, as with all other automatic controls. Security requirements may also demand a general override control to cover emergency conditions at night.

Occupancy detectors are used to detect the presence of people and to control the lighting accordingly. These can rely upon acoustic, infrared, microwave, or other methods of detection. A time lag must normally be built into the system to prevent premature switch-offs or excessive switching.

Depending upon the size of area and number of occupants, it is desirable to provide a degree of individual control that enables personal choice of lighting conditions. In cellular offices, this could be from a combination of high frequency ballasts controlled by a potentiometer or suitable infrared transmitter, which can be used to select or raise and lower the lighting levels. In larger offices, local controls should not noticeably affect the lighting conditions in, and viewing conditions from, adjoining areas (see sections 3.5.2, Localised lighting, and 3.5.3, Local lighting).

Management control systems can address every luminaire in order to programme the appropriate lighting in individual areas. The main advantage to this system is that office alterations can be made and the lighting simply adjusted via the computer to suit the new layout. Combined with local override control, changes can be made without the need for expensive relocation of luminaires and alteration to switching arrangements.

It is possible to interface between a building energy management system (BEMS) and a lighting energy management system (LEMS) in order to provide certain control commands from the BEMS to the lighting. It is not generally cost-effective to use the BEMS to provide discrete localised lighting control to individual luminaires, but rather to achieve load shedding or zone switching.

3.7.1.5 Maintenance control

Through the LEMS it is possible to check the status of the primary and emergency lighting. The system may be programmed to provide the check at prescribed times automatically, the status of each luminaire being checked and recorded.

3.7.2 Human factors

Control systems that are obtrusive (see section 3.8.1.5, Photoelectric control) are counterproductive and may even be sabotaged by the staff. For this reason, dimmer systems are often preferred. Photocells and other sensing circuitry must incorporate a delay to prevent sporadic and disruptive switch-offs, while still responding immediately when a switch-on is called for.

Any control system must ensure that acceptable lighting conditions are always provided for the occupants. Safety and visual effectiveness and comfort must take priority over energy saving.

The Lighting Industry Federation publishes an *Applications Guide*, which deals with this subject matter in further detail.

3.8 Detailed planning

When the overall design has been resolved in general terms, detailed calculations are required to determine such things as the number of luminaires, the glare index, the final cost and so on. When the design has been completed, a check should be made to see how well the original objectives have been met. If this shows that the design is unsatisfactory in some regard, the only course of action is to revise the design until a suitable solution is found. This iterative procedure is a normal part of the design process.

The main calculations that may have to be carried out during the design process are detailed in the following sections of:

- Costs and energy use
- Maintained illuminance
- Average illuminance
- Specification and interpretation of illuminance variation
- Discomfort glare
- Emergency lighting.

3.8.1 Costs and energy use

The most powerful constraints on any design are financial – how much will the scheme cost to install and operate?

Initially it is necessary to establish realistic economic and energy budgets commensurate with the design objectives. At all stages of the design, capital costs and running costs must be scrutinised and controlled. The economics and energy use of the lighting system must be considered within the total building energy use.

3.8.1.1 Financial evaluation

The methods of financial assessment employed by the designer must be acceptable to the client. This can cause difficulties, because grants, tax benefits, tariffs, accounting methods and other factors can vary.

Comparisons are often made with an existing scheme or an alternative design. If the comparison is to be meaningful, the schemes must be designed to equitable standards. The principles of several methods of financial evaluation are discussed in Financial appraisals (see CD).

3.8.1.2 Energy and tariffs

Although many electricity users are able to negotiate a price with an electricity supplier, other consumers will use one of the tariffs published by the regional electricity companies (RECs). The most common commercial and industrial payment systems fall into two categories, quarterly and monthly.

Quarterly tariffs are applicable to most domestic, commercial and small industrial customers. They are relatively simple in structure, comprising a standing charge and one or more unit rates. As an alternative to the standard tariffs, customers may opt for a day/night tariff with different day and night rates but a higher standing charge.

Monthly tariffs are more complex and are generally applicable for large supplies. The most widespread is the maximum demand (MD) tariff, which typically comprises a standing charge, an availability charge linked to the capacity of the supply required, maximum demand charges in the winter months, and one or two unit rates. Also available is a seasonal time of day (STOD) tariff, which may have up to six rates but no MD charges. The RECs will offer advice on the most appropriate tariff for specific applications.

Control of the lighting load profile by switching or dimming, so that unnecessary lighting is not used, will reduce the units consumed. Maximum demand often occurs in the middle of the day, when daylight is available, and MD charges can be reduced if it is possible to shed lighting load at such times. Conversely, it is often possible to add all-night security lighting without increasing the daytime maximum demand, incurring only the appropriate unit cost.

3.8.1.3 Energy use

Designers should ensure that their designs do not waste energy. However, the most important consideration about energy consumption is usually financial. Few users are willing to invest extra money to achieve energy savings unless the savings offer a reasonable rate of return on that investment.

If the design objectives call for particular conditions to be created, these should be provided. If they are not provided, then although the design may use less energy it will not be effective and cannot, therefore, be regarded as satisfactory.

Section 2.4, Energy efficiency recommendations, gives ranges of installed power densities appropriate for various applications. These effectively set limits to the installed load, but other means are required to control energy use and improve operational efficiency.

The load factor for a lighting installation, during a specified period of time, is the ratio of the energy actually consumed to the energy that would have been consumed had the full connected load been operated throughout the specified period. Thus if 25 per cent of the lights in an installation are switched off on average throughout the working day, the load factor will be 0.75. For many installations the load factor will be determined by the ability of the lighting control system to switch the lighting in response to daylight availability. To compare the effectiveness of alternative control systems, the designer will need to estimate the probable annual use of electric lighting under each system.

3.8.1.4 Conventional switching

Field studies of switching behaviour have shown that with traditional switching arrangements, electric lighting is usually either all switched on or all switched off. The act of switching is almost entirely confined to the beginning and end of a period of occupation; people may switch lighting on when entering a room, but seldom turn it off until they all leave. The year-round probability that an occupant will switch lights on when entering a room depends on the time of day, the orientation of the windows, and the minimum orientation-weighted daylight factor on the working area (see Figure 3.12). Daylight factor calculations are covered in section 3.4, Daylight, and *BRE Digest 303: Estimating Daylight in Buildings*. When these are to be used with Figure 3.12, the following orientation weighting factors should be included:

— North-facing windows: 0.77

— East-facing windows: 1.04

— South-facing windows: 1.2

— West-facing windows: 1.00.

For example, if the minimum orientation-weighted daylight factor is 0.6 per cent and work starts at 0900 h, Figure 3.11 shows a 56 per cent probability of switching. If the room is continually occupied, even through the lunch hour, it may be concluded that this same figure, 56 per cent, is the probability that the lights will be on at any moment during the working day. Thus for a lighting installation with a load of 3 kW and a working year

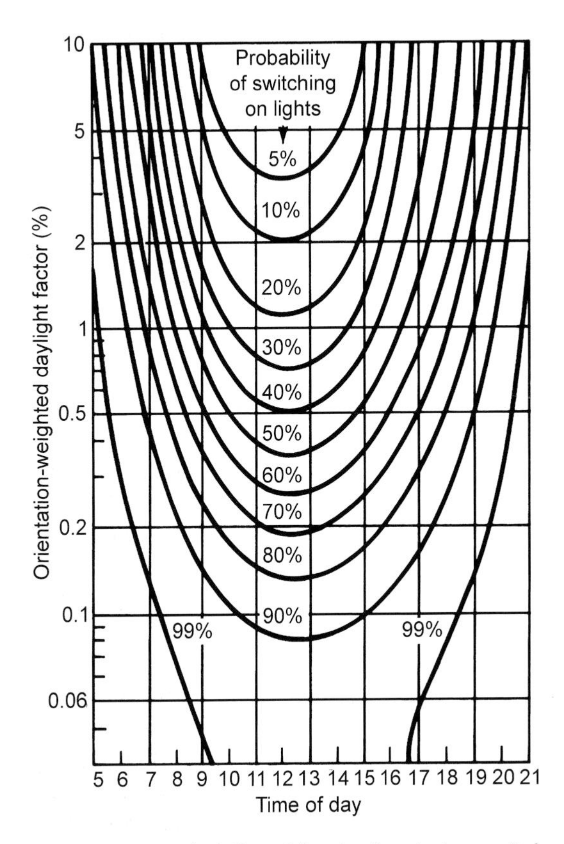

Figure 3.12 Probability of luminaires being switched on

consisting of 260 days, each of 8 hours, the total annual energy consumption would be:

$$260 \times (8 \times 0.56) \times 3 = 3494\,\mathrm{kW/h}$$

In rooms that empty for lunch, the lighting may be switched off by the last person to leave. Here it would be reasonable to treat the periods before and after lunch separately. In the example cited above, if lunch ends at 1330 h, the probability of lights being switched on after lunch is taken as 37 per cent, during lunch as 0 per cent, and during the morning as 56 per cent (as before). Thus if the morning is assumed to be 4 hours and the afternoon 3 hours, with a 1-hour lunch period, the total annual energy consumption will be:

$$260 \times [(4 \times 0.56) + (3 \times 0.37)] \times 3 = 2613\,\mathrm{kWh}$$

This represents a saving of 881 kW/h per year compared with the installation without lunch-time switch-off.

If luminaires are logically zoned with respect to the natural lighting, with convenient pull-cord switches for the occupants to use, each zone can be treated as a separate room. The probability of switching would differ from zone to zone, depending on the minimum orientation-weighted daylight factor in each zone. Figure 3.12 would still be applicable, but the minimum orientation-weighted daylight factor, and consequent energy savings, must be estimated separately for each zone.

A room occupied intermittently can be treated similarly, but some assumption must be made about the periods when the space will be empty.

3.8.1.5 Photo-electric control

On/off switching (Figure 3.13)

Photo-electric controls will normally be zoned to take full advantage of daylight. Figure 3.13 shows the percentage of a normal working year during which the luminaires would be off, as a function of the orientation–weighted daylight factor (see section 3.4.3, Daylight for task illumination) and of the illuminance at which the luminaires are switched; the 'trigger' illuminance. These curves

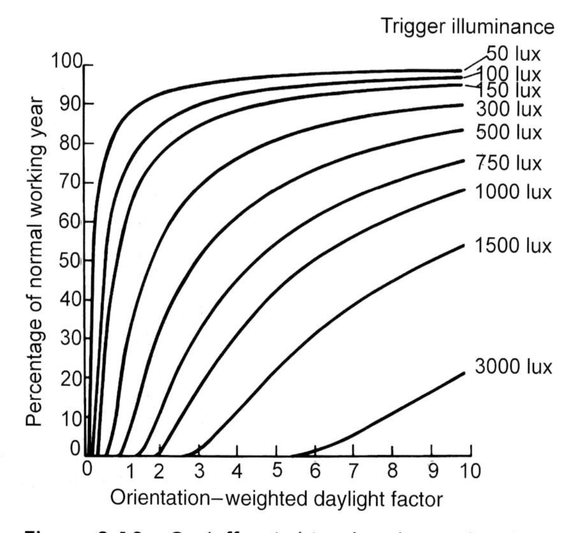

Figure 3.13 On/off switching by photo-electric switching

assume that 'on' and 'off' switching will occur at the same illuminance. Where this is not the case, where the luminaires are switched off at an illuminance appreciably greater than that at which they are switched on, the mean of the two illuminances should be taken as the 'trigger' illuminance.

Dimming 'top-up' (Figure 3.14)

Estimation of energy saving from continuous dimming is complicated by the fact that the lamp circuit luminous efficacy generally decreases as a lamp is dimmed. For a well-designed tubular fluorescent dimming circuit, the cathode heaters consume some 12 per cent of the nominal power consumption and the remaining wattage is roughly proportional to the light output. Figure 3.13 has been constructed on this basis. It shows the percentage of a normal working year during which the luminaires would have to be switched off in order to ensure the energy saving obtainable by continuous photo-electric dimming. It applies to dimmer systems that can control down to 10 per cent output or less.

Dimmer systems can also be designed to operate at less than 100 per cent output when the installation is new so that the maintained illuminance is achieved as a constant value (see sections 3.4, Daylight, and 3.8.2, Maintained illuminance) through life. For example, if the maintenance factor (MF) is 0.5, when new the luminaire will operate at 50 per cent output (i.e. bringing the initial illuminance of 1000 lux down to the design maintained illuminance of 500 lux) and then increase power as the light losses increase. This approach will lead to valuable energy savings even before daylight is taken into account. Over lamp life, the percentage energy saving S_m (relative to an undimmed, fully lit installation) is of the order of:

$$S_m = [(1 - \text{MF})/2] \times 100\%$$

The effect of the maintenance factor is to reduce the dimming range. In the example above, the effective dimming range with new lamps would be from 50 per cent down to, say, 10 per cent (e.g. from a maintained illuminance of 500 lux down to 100 lux). At the end of the effective lamp life, i.e. when the circuit is at full

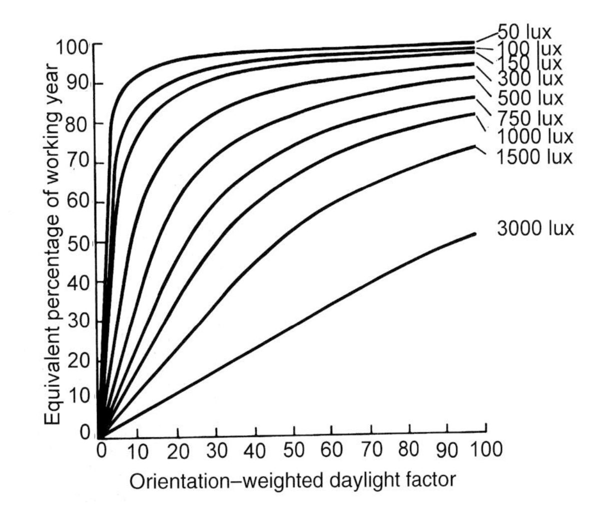

Figure 3.14 Dimming 'top-up' by photo-electric control

power, the dimming range will have increased to 100 – 10 per cent. In the example, this would be 500 – 50 lux.

If the switching is to be relatively unnoticeable to the occupants, the proportion of the electric lighting should not be more than 20 per cent of the total task illuminance. The use of dimming to compensate for the maintenance factor does not, therefore, affect the choice of trigger illuminance, which in the above example would be set at not less than 500 lux.

3.8.2 Maintained illuminance

See also Maintenance of lighting installations (see CD).

Maintained illuminance (E_m) is defined as the average illuminance over the reference surface at the time maintenance has to be carried out by replacing lamps and/or cleaning the equipment and room surfaces.

Before the 1994 edition of this *Code*, in the UK the maintenance factor only took account of losses due to dirt collecting on the lamps, luminaires and room surfaces; it did not include lamp lumen maintenance and lamp failure losses. The new definition of maintenance factor is 'the ratio of maintained illuminance to initial illuminance', i.e. taking account of all losses including lamp lumen maintenance (see Part 4, Glossary, for definitions of the various terms involved).

The maintenance factor (MF) is a multiple of factors:

$$MF = LLMF \times LSF \times LMF \times RSMF$$

where LLMF is the lamp lumen maintenance factor; LSF is the lamp survival factor (used only if spot-replacement of lamps is not carried out); LMF is the luminaire maintenance factor; RSMF is the room surface maintenance factor.

Each of these terms is dealt with more fully in the following sections.

3.8.2.1 Lamp lumen maintenance factor and survival factor

(See Lamps in the LIF Lamp guide, and Lamp replacement – see CD.)

Lamp lumen (luminous flux) maintenance factor (LLMF)

The lumen output from all lamp types reduces with time of operation. The rate of fall-off varies for different lamp types, and it is essential to consult manufacturers' data. From such data it is possible to obtain the lamp lumen maintenance factor for a specific number of hours of operation. The lamp lumen maintenance factor is therefore the proportion of the initial light output that is produced after a specified time, and, where the rate of fall-off is regular, may be quoted as a percentage reduction per thousand hours of operation.

Manufacturers' data will normally be based on British Standards test procedures, which specify the ambient temperature in which the lamp will be tested, with a regulated voltage applied to the lamp and, if appropriate, a reference set of control gear. If any of the aspects of the proposed design are unusual, e.g. high ambient temperature, vibration, switching cycle, operating attitude etc., the manufacturer should be made aware of the conditions and will advise if they affect the life and/or light output of the lamp.

Lamp survival factor (LSF)

As with the lamp lumen maintenance factor, it is essential to consult manufacturers' data. These will give the percentage of lamp failures for a specific number of hours operation, which is only applicable where group lamp replacement, without spot replacement, is to be carried out. These data will also be based on assumptions such as switching cycle, supply voltage and control gear. Manufacturers should be made aware of these aspects and should advise if they will affect the lamp life or lamp survival. Typical lumen maintenance and lamp survival data are given in Table 3.4.

Table 3.4 Typical lumen maintenance and lamp survival data

		Typical values of LLMF and LSF										
		Operation time (1000 h)										
		0.1	0.5	1.0	1.5	2.0	4.0	6.0	8.0	10.0	12.0	14.0
Fluorescent multi- and tri-phosphor	LLMF	1	0.98	0.96	0.95	0.94	0.91	0.87	0.86	0.85	0.84	0.83
	LSF	1	1	1	1	1	1	0.99	0.95	0.85	0.75	0.64
Fluorescent halophosphor	LLMF	1	0.97	0.94	0.91	0.89	0.83	0.80	0.78	0.76	0.74	0.72
	LSF	1	1	1	1	1	1	0.99	0.95	0.85	0.75	0.64
Mercury	LLMF	1	0.99	0.97	0.95	0.93	0.87	0.80	0.76	0.72	0.68	0.64
	LSF	1	1	1	1	0.99	0.98	0.97	0.95	0.92	0.88	0.84
High-pressure sodium	LLMF	1	1	0.98	0.97	0.96	0.93	0.91	0.89	0.88	0.87	0.86
	LSF	1	1	1	1	0.99	0.98	0.96	0.94	0.92	0.89	0.85
High-pressure sodium, improved colour	LLMF	1	0.99	0.97	0.95	0.94	0.89	0.84	0.81	0.79	0.78	—
	LSF	1	1	1	0.99	0.98	0.96	0.9	0.79	0.65	0.50	—

3.8.2.2 Luminaire Maintenance Factor (LMF)

(See also Luminaire cleaning interval – see CD.)

Dirt deposited on or in the luminaire will cause a reduction in light output from the luminaire. The rate at which dirt is deposited depends on the construction of the luminaire and on the extent to which dirt is present in the atmosphere, which in turn is related to the nature of the dirt generated in the specific environment. Table 3.5 gives a list of the luminaire categories and of typical locations where the various environmental conditions may be found.

Table 3.6 shows typical changes in light output from a luminaire caused by dirt deposition, for a number of luminaire and environment categories.

Table 3.5 Luminaire categories and a list of typical locations where the various environmental conditions may be found

Category	Description
A	Bare lamp batten
B	Open top reflector (ventilated self-cleaning)
C	Closed top reflector (unventilated)
D	Enclosed (IP2X)
E	Dustproof (IP5X)
F	Indirect uplighter
Environment	**Typical locations**
Clean (C)	Clean rooms, computer centres, electronic assembly, hospitals
Normal (N)	Offices, shops, schools, laboratories, restaurants, warehouses, assembly workshops
Dirty (D)	Steelworks, chemical works, foundries, welding, polishing, woodwork areas

Table 3.6 Typical changes in light output from a luminaire caused by dirt deposition, for a number of luminaire and environment categories

Time between cleaning (years)		**0.5**			**1.0**			**1.5**	
Environment	C	N	D	C	N	D	C	N	D
Luminaire category									
A	0.95	0.92	0.88	0.93	0.89	0.83	0.91	0.87	0.80
B	0.95	0.91	0.88	0.90	0.86	0.83	0.87	0.83	0.79
C	0.93	0.89	0.83	0.89	0.81	0.72	0.84	0.74	0.64
D	0.92	0.87	0.83	0.88	0.82	0.77	0.85	0.79	0.73
E	0.96	0.93	0.91	0.94	0.90	0.86	0.92	0.88	0.83
F	0.92	0.89	0.85	0.86	0.81	0.74	0.81	0.73	0.65
Time between cleaning (years)		**2.0**			**2.5**			**3.0**	
Environment	C	N	D	C	N	D	C	N	D
Luminaire category									
A	0.89	0.84	0.78	0.87	0.82	0.75	0.85	0.79	0.80
B	0.84	0.80	0.75	0.82	0.76	0.71	0.79	0.74	0.79
C	0.80	0.69	0.59	0.77	0.64	0.54	0.74	0.61	0.64
D	0.83	0.77	0.71	0.81	0.75	0.68	0.79	0.73	0.73
E	0.91	0.86	0.81	0.90	0.85	0.80	0.90	0.84	0.83
F	0.77	0.66	0.57	0.73	0.60	0.51	0.70	0.55	0.65

3.8.2.3 Room surface maintenance factor (RSMF)

Changes in room surface reflectance caused by dirt deposition will cause changes in the illuminance produced by the lighting installation. The magnitude of these changes is governed by the extent of dirt deposition and the importance of inter-reflection to the illuminance produced. Inter-reflection is closely related to the distribution of light from the luminaire, and the room index (K). For luminaires that have a strongly downward distribution, i.e. direct luminaires, inter-reflection has little effect on the illuminance produced on the horizontal working plane. Conversely, indirect lighting is completely dependent on inter-reflections. Most luminaires lie somewhere between these extremes, so most lighting installations are dependent to some extent on inter-reflection.

Table 3.7 shows the typical changes in the illuminance from an installation that occur with time due to dirt deposition on the room surfaces for clean, normal and dirty conditions in small, medium or large rooms lit by direct, semi-direct and indirect luminaires. From this table it is possible to select a room surface maintenance factor appropriate to the circumstances. The areas in which clean, normal and dirty environments are found are given in Table 3.5.

Table 3.7 Typical changes in the illuminance from an installation that occur with time due to dirt deposition on the room surfaces

Time between cleaning (years)			**0.5**			**1.0**			**1.5**	
Room size (K)	Luminaire distribution	C	N	D	C	N	D	C	N	D
Small ($K = 0.7$)	Direct	0.97	0.96	0.95	0.97	0.94	0.93	0.96	0.94	0.92
	Direct/indirect	0.94	0.88	0.84	0.90	0.86	0.82	0.89	0.83	0.80
	Indirect	0.90	0.84	0.80	0.85	0.78	0.73	0.83	0.75	0.69
Medium–large ($K = 2.5–5.0$)	Direct	0.98	0.97	0.96	0.98	0.96	0.95	0.97	0.96	0.95
	Direct/indirect	0.95	0.90	0.86	0.92	0.88	0.85	0.90	0.86	0.83
	Indirect	0.92	0.87	0.83	0.88	0.82	0.77	0.86	0.79	0.74
Time between cleaning (years)			**2.0**			**2.5**			**3.0**	
Room size (K)	Luminaire distribution	C	N	D	C	N	D	C	N	D
Small ($K = 0.7$)	Direct	0.95	0.93	0.90	0.94	0.92	0.89	0.94	0.92	0.88
	Direct/indirect	0.87	0.82	0.78	0.85	0.80	0.75	0.84	0.79	0.74
	Indirect	0.81	0.73	0.66	0.77	0.70	0.62	0.75	0.68	0.59
Medium–large ($K = 2.5–5.0$)	Direct	0.96	0.95	0.94	0.96	0.95	0.94	0.96	0.95	0.94
	Direct/indirect	0.89	0.85	0.81	0.87	0.84	0.79	0.86	0.82	0.78
	Indirect	0.84	0.77	0.70	0.81	0.74	0.67	0.78	0.72	0.64

3.8.3 Average illuminance (lumen method)

The value of design maintained illuminance for an activity or interior is found by referring to sections 2.5, Lighting schedule, and 2.3.2, Illuminance. When this is to be achieved by designing a general lighting installation, the number of luminaires required to achieve the average illuminance can be calculated by means of utilisation factors.

Information on the use of this technique is split into the following sections:

- — Utilisation factors
- — Room index
- — Effective reflectance
- — Maximum spacing-to-height ratio
- — Calculation procedure.

3.8.3.1 Utilisation factors

The utilisation factor UF_S of an installation is the ratio of the total flux received by the reference surface S to the total lamp flux of the installation. The average illuminance E_S over the reference surface S can therefore be calculated from the 'lumen method' formula:

$$E_S = (F \times n \times N \times \text{MF} \times \text{UF}_S)/A_S$$

where F is the initial bare lamp luminous flux (lumens); n is the number of lamps per luminaire; N is the number of luminaires; MF is the maintenance factor (see section 3.8.2, Maintained illuminance); UF_S is the utilisation factor for the reference surface S; A_S is the area of the reference surface S (in m^2).

The formula can be re-arranged to permit the calculation of the number of luminaires required to achieve a chosen illuminance (see section 3.8.3.5, Calculation procedure).

Utilisation factors can be determined for any surface or layout of luminaires, but in practice are only calculated for general lighting systems with regular arrays of luminaires and for the three main room surfaces – the ceiling cavity, the walls, and the floor cavity or horizontal reference plane (see Figure 3.15). Utilisation factors for these surfaces are designated UF_C, UF_W and UF_F, respectively. The method for calculating utilisation factors for these surfaces is given in CIBSE *TM5*.

Although utilisation factors can be calculated by the lighting designer, most manufacturers publish utilisation factors for standard conditions for their luminaires. CIBSE *TM5* defines a standard method of presentation, and states the assumptions on which the tabulated values are based. Table 3.2 is an example of the standard presentation (see also section 3.8.3.4, Maximum spacing-to-height ratio (SHR_{max})). It should be noted that the calculation of UF as described above assumes an empty room; absorption of light by room contents such as furniture and equipment may reduce the achieved illuminance on the working plane.

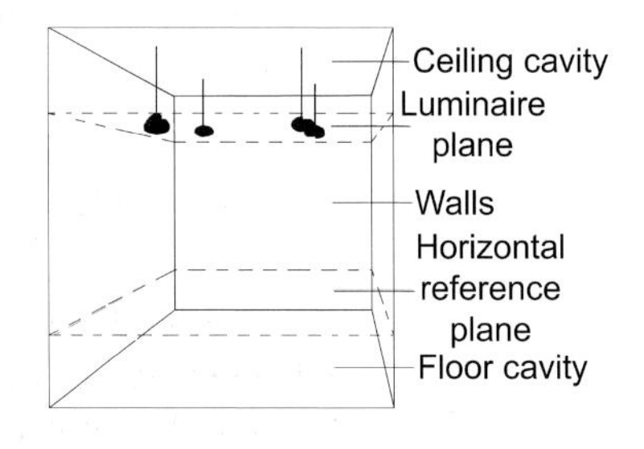

Figure 3.15 Room surfaces

3.8.3.2 Room index

To use utilisation factor tables it is necessary to have values for the room index (K) and the reflectance values of the main room surfaces.

The room index is a measure of the proportions of the room (see Figure 3.15). For rectangular rooms the room index is:

$$K\{(L \times W)/(L + W)h_{\mathrm{m}}\}$$

where L is the length of the room; W is the width of the room; h_{m} is the height of the luminaire plane above the horizontal reference plane. Only if the luminaires are recessed and the working plane is on the floor is h_{m} the floor to ceiling height.

Results may be rounded to the nearest value in the utilisation factor table. If the room is re-entrant in shape (e.g. L-shaped), then it must be divided into two or more non re-entrant sections that are treated separately.

When large areas are subdivided by screens, partitions, structural elements, furniture or machinery that project about the working plane, it is usually advisable to calculate K for the smaller enclosed areas (see sections 1.4, Variation in lighting, and 3.8.4, Specification and interpretation of illuminance variation).

3.8.3.3 Effective reflectance

In order to use utilisation factor tables correctly, the effective reflectances of the ceiling cavity, walls and floor cavity must be calculated.

For the ceiling and floor cavities the cavity indices $\mathrm{CI_C}$ and $\mathrm{CI_F}$ must be calculated. The cavity index (CI), which is similar in concept to the room index, is given by the following:

$$\mathrm{CI} = \frac{(\text{Mouth area of cavity} + \text{Base area of cavity})}{\text{Wall area of cavity}}$$

For rectangular rooms:

$$\mathrm{CI_C} \quad \text{or} \quad \mathrm{CI_F} = LW/(L + W)h = (K \times h_{\mathrm{m}})/h$$

where h is the depth of the cavity.

The effective reflectance $\mathrm{RE_X}$ of the cavity X can then be determined from Table 3.8, or from the simplified, but less accurate, formula:

$$\mathrm{RE_X} = (\mathrm{CI_X} + \mathrm{RA_X})/\{\mathrm{CI_X} + 2(1 - \mathrm{RA_X})\}$$

where $\mathrm{RA_X}$ is the area-weighted average reflectance of the cavity X, and $\mathrm{CI_X}$ is the cavity index of the cavity X.

The average reflectance R_{X} of a series of surfaces S_1 to S_n with reflectances RS_n and areas A_1 to A_n, respectively, is given by:

$$\mathrm{RA_X} = \frac{\sum_{y=1}^{n} R(Sy)A(y)}{\sum_{y=1}^{n} A(y)}$$

It should be noted that in order to calculate the effective reflectances, it is not necessary to know the colours of the surfaces, only the value of reflectance (see SLL publication *Lighting Guide 11: Surface Reflectance and Colour*).

Table 3.8 Cavity index

Reflectance		Cavity index									
Wall	Base	1	2	3	4	5	6	7	8	9	10
0.1	0.1	0.037	0.056	0.067	0.073	0.078	0.081	0.083	0.085	0.087	0.088
	0.2	0.056	0.098	0.122	0.137	0.148	0.155	0.161	0.165	0.168	0.171
	0.3	0.075	0.140	0.178	0.201	0.218	0.229	0.238	0.245	0.250	0.255
	0.4	0.094	0.182	0.234	0.266	0.288	0.303	0.315	0.325	0.332	0.338
	0.5	0.113	0.224	0.289	0.330	0.358	0.378	0.393	0.404	0.414	0.422
	0.6	0.132	0.267	0.345	0.395	0.428	0.452	0.470	0.484	0.496	0.505
	0.7	0.151	0.309	0.401	0.459	0.498	0.527	0.548	0.565	0.578	0.589
	0.8	0.171	0.352	0.458	0.524	0.569	0.601	0.629	0.645	0.660	0.673
Wall	Base	1	2	3	4	5	6	7	8	9	10
0.2	0.1	0.058	0.073	0.080	0.084	0.087	0.089	0.090	0.092	0.092	0.093
	0.2	0.079	0.117	0.137	0.150	0.158	0.164	0.169	0.172	0.175	0.177
	0.3	0.100	0.161	0.195	0.216	0.230	0.240	0.247	0.253	0.258	0.262
	0.4	0.121	0.206	0.253	0.282	0.301	0.316	0.326	0.334	0.341	0.346
	0.5	0.142	0.250	0.311	0.348	0.373	0.391	0.405	0.416	0.424	0.431
	0.6	0.163	0.296	0.369	0.415	0.446	0.468	0.484	0.497	0.507	0.516
	0.7	0.185	0.341	0.428	0.482	0.518	0.544	0.563	0.579	0.591	0.601
	0.8	0.207	0.386	0.487	0.549	0.591	0.620	0.643	0.660	0.674	0.686
Wall	Base	1	2	3	4	5	6	7	8	9	10
0.3	0.1	0.082	0.091	0.094	0.095	0.096	0.097	0.098	0.098	0.098	0.098
	0.2	0.105	0.137	0.153	0.163	0.169	0.174	0.177	0.180	0.182	0.184
	0.3	0.128	0.184	0.213	0.231	0.242	0.251	0.257	0.262	0.266	0.269
	0.4	0.151	0.231	0.273	0.299	0.316	0.328	0.337	0.344	0.350	0.355
	0.5	0.175	0.278	0.334	0.367	0.390	0.406	0.418	0.427	0.434	0.440
	0.6	0.199	0.326	0.395	0.436	0.464	0.484	0.498	0.510	0.519	0.526
	0.7	0.223	0.375	0.456	0.506	0.539	0.562	0.579	0.593	0.604	0.613
	0.8	0.248	0.424	0.518	0.575	0.613	0.641	0.661	0.676	0.689	0.699
Wall	Base	1	2	3	4	5	6	7	8	9	10
0.4	0.1	0.107	0.109	0.108	0.107	0.106	0.105	0.105	0.104	0.104	0.104
	0.2	0.133	0.158	0.170	0.176	0.181	0.184	0.186	0.187	0.189	0.190
	0.3	0.159	0.208	0.232	0.246	0.255	0.262	0.267	0.271	0.274	0.276
	0.4	0.185	0.258	0.295	0.316	0.331	0.341	0.349	0.354	0.359	0.363
	0.5	0.211	0.308	0.358	0.387	0.407	0.420	0.431	0.439	0.445	0.450
	0.6	0.239	0.360	0.422	0.459	0.483	0.500	0.513	0.523	0.531	0.537
	0.7	0.266	0.412	0.486	0.531	0.560	0.581	0.596	0.608	0.617	0.625
	0.8	0.294	0.464	0.552	0.603	0.637	0.661	0.679	0.693	0.704	0.713
Wall	Base	1	2	3	4	5	6	7	8	9	10
0.5	0.1	0.136	0.129	0.123	0.119	0.116	0.114	0.112	0.111	0.110	0.109
	0.2	0.164	0.181	0.187	0.190	0.192	0.194	0.195	0.195	0.196	0.196
	0.3	0.193	0.233	0.252	0.262	0.269	0.274	0.277	0.280	0.282	0.284
	0.4	0.233	0.287	0.317	0.335	0.346	0.354	0.360	0.365	0.369	0.372
	0.5	0.253	0.341	0.383	0.408	0.424	0.436	0.444	0.450	0.456	0.460
	0.6	0.284	0.396	0.450	0.482	0.503	0.517	0.528	0.537	0.543	0.548
	0.7	0.316	0.452	0.518	0.557	0.582	0.600	0.613	0.623	0.631	0.637
	0.8	0.384	0.509	0.587	0.633	0.662	0.683	0.698	0.710	0.719	0.727

Wall	Base	1	2	3	4	5	6	7	8	9	10
0.6	0.1	0.168	0.151	0.139	0.131	0.126	0.123	0.120	0.118	0.116	0.115
	0.2	0.200	0.205	0.205	0.205	0.204	0.204	0.204	0.203	0.203	0.203
	0.3	0.232	0.261	0.273	0.279	0.283	0.286	0.288	0.289	0.290	0.291
	0.4	0.266	0.318	0.341	0.354	0.362	0.368	0.372	0.376	0.378	0.380
	0.5	0.301	0.376	0.410	0.430	0.443	0.451	0.458	0.463	0.467	0.470
	0.6	0.336	0.435	0.481	0.507	0.524	0.535	0.544	0.550	0.556	0.560
	0.7	0.373	0.496	0.552	0.585	0.605	0.620	0.630	0.639	0.645	0.650
	0.8	0.411	0.557	0.625	0.663	0.688	0.705	0.718	0.727	0.735	0.741

Wall	Base	1	2	3	4	5	6	7	8	9	10
0.7	0.1	0.204	0.173	0.155	0.144	0.137	0.132	0.128	0.125	0.122	0.120
	0.2	0.240	0.231	0.224	0.220	0.217	0.215	0.213	0.211	0.210	0.210
	0.3	0.277	0.291	0.295	0.296	0.297	0.298	0.298	0.299	0.299	0.299
	0.4	0.315	0.352	0.366	0.374	0.379	0.382	0.385	0.387	0.388	0.389
	0.5	0.355	0.414	0.439	0.453	0.461	0.468	0.472	0.475	0.478	0.480
	0.6	0.397	0.478	0.513	0.533	0.545	0.554	0.560	0.565	0.568	0.571
	0.7	0.440	0.543	0.589	0.614	0.630	0.641	0.649	0.655	0.660	0.663
	0.8	0.485	0.611	0.666	0.696	0.715	0.728	0.738	0.745	0.751	0.756

Wall	Base	1	2	3	4	5	6	7	8	9	10
0.8	0.1	0.245	0.198	0.173	0.158	0.148	0.141	0.136	0.132	0.129	0.126
	0.2	0.285	0.260	0.245	0.236	0.230	0.225	0.222	0.220	0.218	0.216
	0.3	0.328	0.323	0.318	0.315	0.312	0.311	0.309	0.308	0.308	0.307
	0.4	0.373	0.388	0.393	0.395	0.396	0.397	0.398	0.398	0.398	0.399
	0.5	0.419	0.456	0.469	0.477	0.481	0.484	0.487	0.488	0.490	0.491
	0.6	0.468	0.525	0.548	0.560	0.567	0.573	0.576	0.579	0.582	0.583
	0.7	0.519	0.596	0.627	0.644	0.655	0.662	0.667	0.671	0.674	0.677
	0.8	0.573	0.670	0.709	0.730	0.743	0.752	0.759	0.764	0.768	0.771

3.8.3.4 Maximum spacing-to-height ratio

The maximum spacing between the centres of luminaires divided by the mounting height above the horizontal reference plane should not exceed the maximum spacing-to-height ratio (SHR_{max}) if uniformity of illuminance is to be acceptable for general lighting. The SHR_{max} for the luminaire can be obtained using the method described in CIBSE *TM5*. The utilisation factor table is calculated for the nominal spacing-to-height ratio (SHR_{nom}). This is the SHR in the series 0.5, 0.75, 1.0 etc. that is not greater than SHR_{max}. For linear luminaires with conventional distributions, the SHR_{max} can be supplemented by the maximum transverse spacing-to-height ratio ($SHR_{max\ tr}$) for continuous lines of luminaires.

The axial spacing-to-height ratio (SHR_{ax}) should not exceed SHR_{max}, and the transverse spacing-to-height ratio SHR_{tr} should not exceed the maximum transverse spacing-to-height ratio $SHR_{max\ tr}$. In addition, the product of SHR_{ax} and SHR_{tr} should not exceed $(SHR_{max})^2$. Thus:

$$SHR_{ax}SHR_{tr} \leq (SHR_{max})^2, \quad \text{and}$$

$$SHR_{ax} \leq SHR_{max}, \quad \text{and}$$

$$SHR_{tr} \leq SHR_{max\ tr}$$

For some luminaires, notably those with a rapid rate of change in their intensity distributions, extra spacing-to-height ratio information may be given. This can be provided in the form of a graph showing acceptable combinations of axial and transverse spacing.

The mid-area is defined for a 4×4 array of luminaires as the rectangular (or square) area on the horizontal plane with corners directly below the centres of the four central luminaires. The mid-area ratio (MAR) is the ratio of the minimum direct illuminance to the maximum direct illuminance calculated over the mid-area. This is obtained by calculating the direct illuminance at every point on a regular 9×9 grid with the corner points of the grid at the corners of the area, i.e. one under the centre of each of the four luminaires. This is the general method described in *TM5* and is now preferred to the simpler mid-point ratio (MPR) method, which is also included in *TM5*.

These spacing rules are provided as a guide for general design work, and give approximate advice. Where computer programs are available, these can be used to check that correct uniformity has been achieved over task areas.

The SHR data and the majority of lighting programs take no account of obstructions or obstruction losses, or of shadows or modelling. **Designers should not design installations close to their maximum spacing without considering the implications of doing so in terms of shadows, modelling, obstructions and the effect of lamp failures or local switching.**

Note: The use of SHR to control uniformity is only applicable when using regular arrays of luminaires.

3.8.3.5 Calculation procedure

The following procedure gives guidance on the sequence of calculations to be performed when calculating the number of luminaires necessary to obtain a chosen average illuminance on the horizontal reference plane by the lumen method.

(*a*) Calculate the room index K, the floor cavity index CI_F and the ceiling cavity index CF_C (see sections 3.8.3.2, Room index, and 3.8.3.3, Effective reflectance).

(*b*) Calculate the effective reflectances of the ceiling cavity, walls and floor cavity. Remember to include the effect of desks or machines in the latter (see section 3.8.3.3, Effective reflectance).

(*c*) Determine the utilisation factor value from the manufacturer's data for the luminaire, using the room index and effective reflectances calculated as above. Apply any correction factors (given in the utilisation factor table) for lamp type or mounting position to the utilisation factor (UF) value.

(*d*) Determine the maintenance factor (see section 3.8.2, Maintained illuminance).

(*e*) Insert the appropriate variables into the lumen method formula to obtain the number of luminaires required:

$$N = \frac{E_F \times A_F}{F \times n \times \text{MF} \times \text{UF}_F}$$

where E_F is the average illuminance to be provided on the working plane (lux); A_F is the area of the working plane (m^2); F is the initial bare lamp luminous flux (lumens); n is the number of lamps per luminaires; MF is the maintenance factor; UF_F is the utilisation factor for the plane; F refers to the horizontal reference plane.

(*f*) Determine a suitable layout.

(*g*) Check that the geometric mean spacing-to-height ratio of the layout is within the range of the nominal spacing-to-height ratio (SHR_{nom}) for which the utilisation factor table is based, i.e.

$$\sqrt{SHR_{ax} \times SHR_{tr}} = SHR_{nom} \pm 0.5$$

If this is not the case, the UF can be recalculated for the actual spacing using the method given in *TM5*. Corrections are also given for luminaire-to-wall spacing other than half of the spacing between luminaires.

(*h*) Check that the proposed layout does not exceed the maximum spacing-to-height ratios (see section 3.8.3.4, Maximum spacing-to-height ratio).

(*i*) Calculate the illuminance that will be achieved by the final layout (see section 3.8.3.1, Utilisation factors).

3.8.4 Specification and interpretation of illuminance variation

It is possible to describe illuminance variation as a series of values or as some form of graphical representation (plot) of a magnitude of variation over a surface. The main methods used in this *Code* are 'uniformity', which is concerned with illuminance conditions on the task and immediate surround, and 'diversity', which expresses changes in illuminance across a larger space. Values of both may be calculated or measured from a grid of discrete illuminance values over a surface (see Figure 3.16 and the figure in Field surveys (see CD)).

The illuminance conditions on the task area locations across the working plane may be represented by a grid of points over the task and immediate surround. If a number of task areas exist, the uniformity must be determined for each and the worst taken as the limiting value of uniformity for the installation. The concept

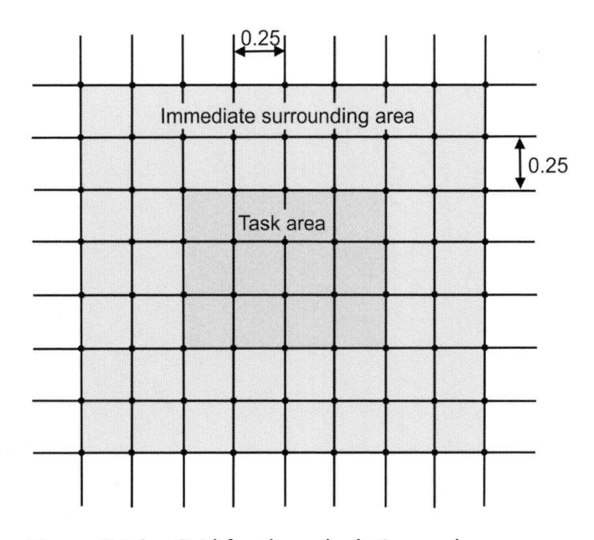

Figure 3.16 Grid for the calculation and measurement of uniformity for a task area and immediate surround

of uniformity thus does not apply to the whole working plane, but to a series of defined task areas on the working plane. In the calculation of illuminance diversity, grid points within 0.5 m of walls or large fixed obstructions are ignored (see Figure 3.17). This is because planar illuminances will (under normal circumstances) fall around the room perimeter and near large obstructions, such as partitions or large structural columns, that project above the working plane.

Calculation or measurement of illuminance in these positions will therefore have little practical value, and particular care should be taken in interpretation of computer-generated illuminance grids to make sure that the grid values chosen for assessment of illuminance variation are not at the very edge of rooms or adjacent to obstructions (see section 1.4, Variation in lighting).

Diversity is a measure of the range of lighting over a specified plane, normally the horizontal working plane, in the space. It is meant to limit the peaks and troughs in the levels of lighting seen by the users of that space. If the range or diversity of illuminance is too wide, then the space is likely to be viewed as disconcerting or distracting by most users. If the range of illuminance is too low, then the space may be seen as bland or uninteresting. Diversity is the ratio of minimum illuminance to maximum illuminance found over the working plane of the main area of a room or space, and should not exceed 1 : 5.

3.8.4.1 Calculation of illuminance variation

No design method is available that will enable the various recommendations relating to illuminance uniformity or diversity to be optimised within a particular proposed design solution.

Widespread use is now made of computer programs that are capable of undertaking analysis, in considerable detail, of illuminance conditions in proposed installations. Most programs are capable of calculating illuminance on a grid of points across a

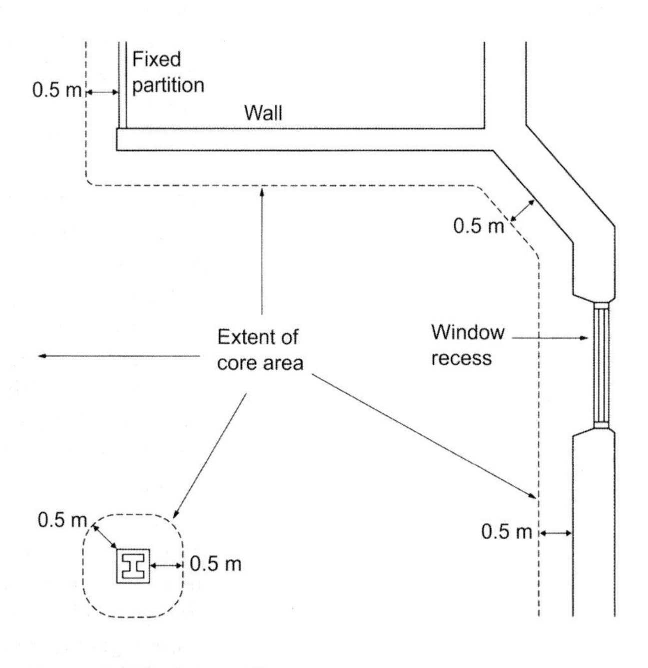

Figure 3.17 Core office space

working plane, and in some cases other room surfaces and illuminance variation quantities are commonly generated as part of the output of such programs.

The purpose of this section is to explain how to calculate the different variation criteria to confirm that the design objectives have been met. To calculate diversity, illuminance values should be calculated on a grid of points located symmetrically over the core area of the working plane. For installations with ceiling heights of up to 5 m lit by a regular array of ceiling-mounted luminaires, the grid of points should normally be at a spacing of 1 m. For other types of installation the grid size may vary. In larger interiors lit by luminaires with a smooth medium-to-wide intensity distribution and those with a mounting height greater than 5 m, the calculation grid size may be increased, in this case the total number of calculation points being determined by reference to the table in Field surveys (see CD). On the other hand, a calculation grid size of less than 1 m may be necessary for installations where abrupt variations in working plane illuminance may occur – for example, those using luminaires with bat-wing or narrow distribution. Care must be taken to ensure that the luminaire and calculation grids do not coincide, and this may also necessitate a small change in the size of the calculation grid. The calculated illuminance value at each point must be made up of both illuminance arriving directly from light sources and illuminance received at the point after reflection from room surfaces (see CD for Calculations guide). The illuminance diversity is calculated from the maximum and minimum illuminance at any point on the grid over the core area of the working plane, ignoring calculation points within 0.5 m of obstructions (see section 1.4, Variation in lighting).

The calculation procedure for uniformity differs slightly depending upon whether or not the size and position of the task areas on the working plane are known. If task location can be identified, then the illuminance calculation should be on a 0.25-m^2 grid over the task and immediate surround for a typical workstation. If the task locations are unknown and may be at any position on the working plane, a suitable number of grid points used for the illuminance diversity assessment are chosen. These should at least include the points of minimum and maximum illuminance over the coarse grid. In this case, the task area may be assumed to be a 0.5-m square with one corner coinciding with the coarse grid point. The illuminance is then calculated on nine points of a 0.25-m grid and at points in a band 0.5 m wide surrounding the area, to check that the illuminance of the surround is at least the level given in Table 2.1. Task uniformity is assessed using the area-weighted arithmetic average of the points within the individual central task areas and the minimum point illuminance value within each area. The lowest value of task uniformity calculated for the various potential or actual locations is taken as representative of the whole installation (see section 1.4, Variation in lighting).

Any discrete set of grid points (measured or calculated) cannot necessarily capture the minimum or maximum values over an area, and under some circumstances abrupt variations in rate of change can be missed by a grid. The representation of illuminance using a grid is therefore an approximation, and is only as good as the choice of location and size of the grid. Hence the recommendation that two grids be used: a series of fine grids over task areas, and a coarse grid across the core area of the working plane (see CD, figure in Field surveys).

3.8.5 Discomfort glare

Freedom from discomfort caused by glare is an important criterion of lighting quality. The sensation of glare in a lighting installation, experienced by people in the form of discomfort, annoyance or irritation, is a complex function of the imbalance within the luminance pattern experienced by the visual mechanism, certain parts (glare sources) lying above the range to which the eye is adapted at the time.

Experimental work has shown that the main factors influencing discomfort glare are the luminance of the sources and their apparent size, their position in the field of view, and the luminance of the general environment. These factors can be combined in a formula to determine the degree of discomfort glare, which is known as the unified glare rating (UGR). The UGR formula may be used to evaluate the glare stimulus within a given environment. However, the actual perception of glare varies from person to person. The calculated index for a particular interior and lighting system can then be compared with a limiting value given in the recommendations in this *Code*. Lighting that is considered to be uncomfortable in one environment may be acceptable in another, and it is possible to determine values that represent (from experience and practice) a suitable comfort criterion that is acceptable to most people for a given occupation in a given location. If the calculated value is greater than the recommended limit, modifications to the lighting system or the interior will be required.

In the Calculations Guide (see CD), the recommended procedure is given for the evaluation of the UGR for a lighting installation from the basic formula. This method has the greatest applicability, being capable of use for any arrangement of glare sources and for any position of observation. Recommendations are given for the determination of each of the parameters in the formula. Whilst the direct application of the this sort of calculation to various points within a space may give insights in to the variation of glare within that space, the limits given in the Lighting schedule (section 2.5) are based on UGR values calculated in standard conditions.

In the Calculations Guide (see CD), recommendations are made for the preparation in a standard form of tables of UGR data to be published for specific luminaires, and the method of determination of the parameters in the formula is included. This work is based on a regular array of luminaires in an overhead installation viewed from a set position.

It is expected that the luminaire manufacturers will prepare and publish these tabulated data as part of the photometric data for each luminaire. For the design of a lighting installation, the data can be converted simply, by means of predetermined terms, to the actual conditions of lamp light output under study.

Example calculations of discomfort glare (see CD) give examples of the calculation of a standard glare table and the application of a glare table to calculate the glare in a sample room.

3.8.6 Emergency lighting

Emergency lighting is provided for use when the main lighting fails for whatever reason. There are two types: escape lighting and standby lighting.

Escape lighting

The Fire Precautions Act 1971 and the Health and Safety at Work etc. Act of 1974 make it obligatory to provide adequate means of escape in all places of work and public resort. Emergency lighting is generally considered to be an essential part of this requirement. *BS 5266 Code of Practice for the Emergency Lighting of Premises* lays down minimum standards for the design, implementation and certification of emergency lighting installations.

Escape lighting is provided to ensure the safe and effective evacuation of the building. It must:

— indicate clearly and unambiguously the escape routes

— illuminate the escape routes to allow safe movement towards and out of the exits

— ensure that fire alarm call points and fire equipment provided along the escape route can be readily located.

See CD for more information on emergency and escape luminaires.

Standby lighting

It may not be possible, or in some cases desirable, to evacuate some building areas immediately in the event of an emergency or power failure. This may be because life would be put at risk, as in a hospital operating theatre, or in some chemical plants where safe shutdown procedures must be used. In shops and offices, for instance, it may be advisable to determine the nature of the emergency before deciding upon evacuation. To evacuate a large store for a simple interruption of the public supply is to risk panic and the loss of stock by opportunist thieves. In these circumstances, standby lighting is required to allow appropriate actions to take place or activities to continue. The level of standby lighting will depend upon the nature of the activities, their duration and the associated risk, and can range from 50–100 per cent of the maintained illuminance according to circumstances.

Standby lighting can be regarded as a special form of conventional lighting and dealt with accordingly. Escape lighting requires different treatment.

This section on emergency lighting is split into the following headings:

— Escape lighting requirements

— Marking the route

— Illuminating the route

— Other important factors

— Systems and calculations

— Maintenance and testing
— Planning sequence.

3.8.6.1 Escape lighting requirements

The *Building Regulations Approved Document B – Fire Safety* defines the types of building and areas within buildings requiring emergency lighting. This document refers to *BS 5266* as the standard to be used for the installation of escape lighting systems.

In addition, The Fire Precautions Act 1971 gives certain powers to local Fire Authorities that allows them to impose standards more onerous than those defined by *BS 5266*, and designers should verify the requirements of the appropriate Fire Authority when *designing* emergency lighting systems.

For certain applications, such as cinemas, theatres, lecture rooms and photographic darkrooms, the level of maintained emergency lighting and the luminance of signs must not adversely affect the primary use of the area.

BS 5266 incorporates, as Part 7, the full requirements of the European Emergency Lighting Standard, *EN 1838.*

3.8.6.2 Marking the route

All exits and emergency exits must have exit or emergency exit signs. Where direct sight of an exit is not possible, or there could be doubt as to the direction, then additional signs are required. These signs should comply with the Safety Signs Directive (*92/58EEC*) and be language-independent pictograms, although some Fire Authorities may require supplementary signs to *BS 5499.*

3.8.6.3 Illuminating the route

BS 5266 Part 1 requires that the centre line of permanently unobstructed escape routes up to 2 m wide should be illuminated to a minimum of 0.2 lux, but preferably to 1.0 lux, at floor level. Additionally, 50 per cent of the 2-m width should be lit to a minimum of 0.1 lux at floor level.

In larger areas where the escape route is not clearly defined, open area 'anti-panic' lighting should be provided, as detailed in *BS 5266* Part 7.

3.8.6.4 Other important factors

Speed of operation: emergency lighting must be provided within 5 s of the failure of the main lighting system. If the occupants are familiar with the building, this time can be increased to 15 s at the discretion of the enforcing authority.
Glare: in order to limit disability glare, luminaires should be placed at least 2 m above floor level. The luminous intensity of emergency luminaires should be controlled as required by *BS 5266* Part 7 Table 1 within a zone 60–90° from the downward vertical for level escape routes, and at all angles for other areas.
Exits and changes of direction: luminaires should be located near each exit door and emergency exit door, and at points where it is necessary to emphasise the position of potential hazards – such as changes of direction, staircases, changes of floor level and so on.

Fire equipment: fire-fighting equipment and fire alarm call points along the escape route must be adequately illuminated at all material times.
Lifts and escalators: although lifts must not be used in the event of an emergency, they should be illuminated. Emergency lighting is required in each lift car in which people can travel. Escalators must be illuminated to the same standard as the escape route to prevent accidents.
Special areas: emergency lighting luminaires are required in all control rooms and plant rooms. In toilets, lobbies and closets exceeding 8 m^2 (or if less than 8 m^2 without borrowed light), escape lighting should be provided as if it were part of an escape route.
High-risk task area lighting: there are certain processes where there would be an increased risk to operators or other people near the task should the general lighting fail – for example, where there is high-speed machinery or exposed flame-heating operations. For such areas, *BS 5266* Part 7 requires that a maintained emergency lighting level of at least 10 per cent of the normal task illuminance be provided, with a minimum value of 15 lux.

3.8.6.5 Systems and calculations

The design aspects of conventional emergency lighting systems are discussed in detail in CIBSE *Technical Memoranda 12: Emergency Lighting.*

3.8.6.6 Maintenance and testing

The regular maintenance of emergency lighting equipment is essential to its correct operation. Generators will require periodic servicing of the prime mover, central batteries will require electrolyte checks, sealed batteries will need checking for loss of capacity, and the luminaires will require checking for correct operation and light output.

A maintenance and testing schedule should be prepared based on the recommendations of the equipment manufacturers and the requirements of *BS 5266* and the appropriate enforcing authority.

If central battery systems are used then it is necessary to follow *BS EN 50171 (2001) – Central Power Systems.*

3.8.6.7 Planning sequence

When planning an emergency lighting system, the following sequence will help.

(*a*) Define the exits and emergency exits.

(*b*) Mark the escape routes.

(*c*) Identify any problem areas, e.g. areas that will contain people unfamiliar with the building, plant rooms, escalators, fire alarm call points, fire equipment etc.

(*d*) Mark the location of exit signs. These can be self-illuminated or illuminated by emergency lighting units nearby. Mark these on the plan.

(*e*) Where direction signs are required, mark these and provide necessary lighting.

(*f*) Identify the areas of the escape route illuminated by the lighting needed for signs.

(*g*) Add extra luminaires to complete the lighting of the escape route, paying attention to stairs and other hazards. Remember to allow for shadows caused by obstructions or bends in the route.

(*h*) Add extra luminaires to satisfy the problem areas identified in item (*c*) of this sequence. Make sure that lighting outside the building is also adequate for safe evacuation.

(*i*) Prepare a schedule for the regular maintenance and testing of all equipment, and ensure that the building owners and operators are familiar with its requirements.

3.9 Design checklist

The designer should check systematically that all the factors relevant to the design of the lighting installation have been taken into account. In the following checklist, the headings indicate the areas to be considered and the most commonly occurring questions. In any specific situation there may be other questions that need to be considered.

Objectives

— Safety requirements: What hazards need to be seen clearly, what form of emergency lighting is needed, is a stroboscopic effect likely?

— Task requirements: Where are the tasks to be performed in the interior, what planes do they occupy? What aspects of lighting are important to the performance of these tasks? Are optical aids necessary?

— Appearance: What impression is the lighting required to create?

Constraints

— Statutory: Are there any statutory requirements that are relevant to the lighting installation?

— Financial: What is the budget available, and what is the relative importance of capital and running costs including maintenance?

— Physical: Is a hostile or hazardous environment present? Are high or low ambient temperatures likely to occur? Is noise from control gear likely to be a problem? Are mounting positions restricted, and is there a limit on luminaire size?

— Historical: Is the choice of equipment restricted by the need to make the installation compatible with existing installations?

Specification

— Source of recommendations: What is the source of the lighting recommendations used? How authoritative is this source?

— Form of recommendations: Have all the relevant lighting variables been considered, e.g. design maintained illuminance, uniformity, illuminance ratios, surface reflectances and colours, light source colour, colour rendering group, limiting glare index, veiling reflections?

- Qualitative requirements: Have the aspects of the design that cannot be quantified been carefully considered?

General planning

- Daylight and electric lighting: What is the relationship between these forms of lighting? Is it possible or desirable to provide a control system to match the electric lighting to the daylight available?
- Protection from solar glare and heat gain: Are the windows designed to limit the effects of solar glare and heat gain on the occupants of the building? Do the window walls have suitable reflectances?
- Choice of electric lighting system: Is general, localised or local lighting for task or display most appropriate for the situation? Does obstruction make some form of local lighting necessary?
- Choice of lamp and luminaire: Does the light source have the required lumen output, luminous efficacy, colour properties, lumen maintenance, life, run-up and re-strike properties? Is the proposed lamp and luminaire package suitable for the application? Is air handling heat recovery appropriate? Will the luminaire be safe in the environmental conditions? Will it withstand the environmental conditions? Does it have suitable maintenance characteristics and mounting facilities? Does it conform to *BS 4533/BS EN 60598-1* or other appropriate standard? Does the luminaire have an appropriate appearance, and will it enable the desired effect to be created? Are reliable photometric data available?
- Maintenance: Has a maintenance schedule been agreed? Has a realistic maintenance factor been estimated based on the agreed schedule or, if not, have the assumptions used to derive the maintenance factor been clearly recorded? Is the equipment resistant to dirt deposition? Can the equipment be easily maintained, is the equipment easily accessible, and will replacement parts be readily available?
- Control systems: Are control systems for matching the operation of the lighting to the availability of daylight and the pattern of occupancy appropriate? Is a dimming facility desirable? Have manual switches or local override facilities been provided, are they easily accessible, and is their relationship to the lighting installation understandable?
- Interactions: How will the lighting installation influence other building services? Is it worth recovering the heat produced by the lamps? If so, have the airflow rates been checked in relation to the operating efficacy of the lamps?

Detailed planning

- Layout: Is the layout of the installation consistent with the objectives and the physical constraints? Has allowance been made for the effects of obstruction by building structure, other services, machinery and furniture? Has the possibility of undesirable high luminance reflections from specular surfaces been considered? Does the layout conform to the spacing-to-height ratio criteria?

- Mounting and electrical supply: How are the luminaires to be fixed to the building? What system of electricity supply is to be used? Does the electrical installation comply with the latest edition (with any amendments) of the *IEE Requirements for Electrical Installations (BS 7671)*?
- Calculations: Have the design maintained illuminance and variation been calculated for appropriate planes? Has an acceptable maintenance programme been specified? Have the most suitable calculation methods been used? Has the glare rating been calculated? Have up-to-date and accurate lamp and luminaire through-life photometric data been used?
- Verification: Does the proposed installation meet the specification of lighting conditions? Is it within the financial budget? Is the power density within the recommended range? Does the installation fulfil the design objectives?

3.10 Statement of assumptions

When submitting a design proposal to clients, it will usually be necessary to supply information on the following topics:

- the design specification, i.e. the type of lighting system, the design maintained illuminance, illuminance variation, the maintenance programme, the glare index, the lamp colour properties, the wall-to-task illuminance ratio, the ceiling-to-task illuminance ratio, and other criteria as applicable
- the equipment to be used, e.g. lamps, luminaires, control systems
- the layout of the equipment
- the costs, in an appropriate form
- the lighting conditions that will be achieved if the maintenance programme is implemented
- the calculation and measurement tolerances (see CD) that apply to these values
- the power density and operating efficacy of the installation
- all assumptions made in the design.

The level at which each of these topics is covered is a matter of commercial judgement. Ambiguity in the information supplied to the client should be avoided – particularly regarding the lighting conditions that will be achieved, the maintenance requirements and the assumptions made in the design. If the client is to compare design proposals on an equitable basis, ideally it is the client (or the client's consultant) who should specify the major design criteria and the assumptions to be made. In any case it is essential that the assumptions made in the design are stated by the designer for each aspect of the lighting conditions. Table 3.9 lists the assumptions that are usually involved in the estimation of the lighting conditions achieved by a general lighting installation. If localised lighting is being proposed, it will also be necessary to state the areas to which each illuminance applies and to give details for each area separately. If local lighting is being proposed, it will be necessary to give details of the general surround illuminance and the task illuminance, the latter being divided into the contributions from the local luminaire and from the general surround lighting. Special situations may involve additional assumptions, in which case these too should be stated.

Table 3.9 Assumptions to be made explicit when describing the lighting conditions that will be produced by a proposed general lighting installation (these assumptions may be made by the designer or by the client and given to the designer in the form of a specification)

Lighting condition	Assumptions that need to be stated
Initial illuminance	Room index, effective reflectance of ceiling cavity, walls and floor cavity used in establishing the utilisation factor; the initial luminous flux of the lamp used. Supply voltage, ambient temperature, obstruction losses etc.
Illuminance at a specified time	As for initial illuminance, plus the elapsed time for which the illuminance is given, and maintenance factor (see below)
Glare rating (UGR)	Calculation method and viewing position
Wall-to-task illuminance ratio	As for initial and maintained illuminance
Ceiling-to-task illuminance ratio	As for initial and maintained illuminance
Vector/scalar ratio	As for initial and maintained illuminance
Maintenance factor	Elapsed time for which maintenance factor is given, environmental conditions, lamp lumen maintenance factor, lamp survival factor and hours of operation of lamps, the luminaire maintenance factor and luminaire cleaning schedule, room surface maintenance factor and room cleaning and painting schedule
Power density	As maintenance factor
Operating efficacy	Maximum hours of use and hours of equivalent full installation use assumed in the calculation of load factor

Part 4 Glossary

The glossary contains definitions of the basic terms and criteria necessary for the specification of lighting.

The definitions closely follow those given in the draft CEN standard Lighting Applications – Basic terms and criteria for specifying lighting requirements.

Absorptance (α): ratio of the luminous flux absorbed in a body to the luminous flux incident on it.

Note: the absorptance generally depends on the direction and spectral distribution of the incident light and on the surface finish.

Accommodation: adjustment of the power of the lens of the eye for the purpose of focusing an image of an object on the retina.

Technically defined as adjustment of the dioptric power of the crystalline lens by which the image of an object, at a given distance, is focused on the retina.

Adaptation: process that takes place as the visual system adjusts to the luminance and colour of the visual field or the final state of this process.

Technically defined as the process by which the state of the visual system is modified by previous and present exposure to stimuli that may have various luminances, spectral distributions and angular subtenses.

Notes: (1) the terms *light adaptation* and *dark adaptation* are also used, the former when the luminances of the stimuli are of at least several candelas per square metre, and the latter when the luminances are of less than some hundredths of a candela per square metre; (2) adaptation to specific spatial frequencies, orientations, sizes etc. are recognised as being included in this definition.

Average illuminance: illuminance averaged over the specified area. Unit: lux.

Note: in practice this may be derived either from the total luminous flux failing on the surface divided by the total area of the surface, or alternatively from an average of the illuminances at a representative number of points on the surface.

Average luminance (L_{av}): luminance averaged over the specified area or solid angle. Unit: candela per square metre.

Ballast: a device connected between the supply and one or more discharge lamps that serves mainly to limit the current of the lamp(s) to the required value.

Note: a ballast may also include means of transforming the supply voltage and correcting the power factor and, either alone or in combination with a starting device, provide the necessary conditions for starting the lamp(s).

Ballast lumen factor: ratio of the luminous flux emitted by a reference lamp when operated with a particular production ballast to the luminous flux emitted by the same lamp when operated with its reference ballast.

Brightness: attribute of the visual sensation associated with the amount of light emitted from a given area. It is subjective correlate of luminance.

Technically defined as luminosity (obsolete): attribute of a visual sensation according to which an area appears to emit more or less light.

Brightness contrast: subjective assessment of the difference in brightness between two or more surfaces seen simultaneously or successively.

Chromaticity: property of a colour stimulus defined by its chromaticity coordinates, or by its dominant or complementary wavelength and purity taken together.

Chromaticity coordinates: ratio of each of a set of three tristimulus values to their sum.

Notes: (1) as the sum of the three chromaticity coordinates equals 1, two of them are sufficient to define a chromaticity; (2) in the CIE standard colorimetric systems, the chromaticity coordinates are presented by the symbols x, y, z and x_{10}, y_{10}, z_{10}.

Colorimeter: instrument for measuring colorimetric quantities, such as the tristimulus values of a colour stimulus.

Colour contrast: subjective assessment of the difference in colour between two or more surfaces seen simultaneously or successively.

Colour rendering (of a light source): effect of a light source on the colour appearance of objects compared with their colour appearance under a reference light source. The definition is more formally expressed as the effect of an illuminant on the colour appearance of objects by conscious or subconscious comparison with their colour appearance under a reference illuminant.

Colour stimulus: visible radiation entering the eye and producing a sensation of colour, either chromatic or achromatic.

Colour temperature (T_C): the temperature of a Planckian (black body) radiator whose radiation has the same chromaticity as that of a given stimulus. Unit: K.

Note: the reciprocal colour temperature is also used, unit K^{-1}.

Contrast: (1) In the perceptual sense, assessment of the difference in appearance of two or more parts of a field seen simultaneously or successively (hence: brightness contrast, lightness contrast, colour contrast, simultaneous contrast, successive contrast, etc.). (2) In the physical sense, quantity intended to correlate with the perceived brightness contrast, usually defined by one of a number of formulae that involve the luminances of the stimuli considered, for example: $\Delta U L$ near the luminance threshold, or L_1/L_2 for much higher luminances.

Correlated colour temperature (T_{CC}): the temperature of the Planckian (black body) radiator whose perceived colour most closely resembles that of a given stimulus at the same brightness and under specified viewing conditions. Unit: K.

Notes: (1) the recommended method of calculating the correlated colour temperature of a stimulus is to determine on a chromaticity diagram the temperature corresponding to the point on the Planckian locus that is intersected by the agreed isotemperature line containing the point representing the stimulus; (2) reciprocal correlated colour temperature is used rather than reciprocal colour temperature whenever correlated colour temperature is appropriate.

Cosine correction: correction of a detector for the influence of the incident direction of the light.
Note: for the ideal detector, the measured illuminance is proportional to the cosine of the angle of incidence of the light, where the angle of incidence is the angle between the direction of the light and the normal to the surface of the detector.

Cut-off: technique used for concealing lamps and surfaces of high luminance from direct view in order to reduce glare.
Note: in public lighting, distinction is made between full-cut-off luminaires, semi-cut-off luminaires and non-cut-off luminaires.

Cut-off angle (of a luminaire): angle, measured up from nadir, between the vertical axis and the first line of sight at which the lamps and the surfaces of high luminance are not visible.

Cylindrical illuminance (at a point) (E_Z): total luminous flux falling on the curved surface of a very small cylinder located at the specified point divided by the curved surface area of the cylinder. The axis of the cylinder is taken to be vertical unless stated otherwise. Unit: lux.
Technically defined by the formula:

$$E_Z = \frac{1}{\pi} \int_{4\pi sr} L \sin \varepsilon d\Omega$$

where $d\Omega$ is the solid angle of each elementary beam passing through the given point, L is its luminance at that point, and ε is the angle between it and the given direction.

Daylight: visible part of global solar radiation.
Note: when dealing with actinic effects of optical radiations, this term is commonly used for radiations extending beyond the visible region of the spectrum.

Daylight factor (D): ratio of the illuminance at a point on a given plane due to the light received directly or indirectly from a sky of assumed or known luminance distribution, to the illuminance on a horizontal plane due to an unobstructed hemisphere of this sky. The contribution of direct sunlight to both illuminances is excluded.
Notes: (1) glazing, dirt effects, etc. are included; (2) when calculating the lighting of interiors, the contribution of direct sunlight must be considered separately.

Diffuse sky radiation: that part of solar radiation which reaches the Earth as a result of being scattered by the air molecules, aerosol particles, cloud particles or other particles.

Diffused lighting: lighting in which the light on the working plane or on an object is not incident predominantly from a particular direction.

Direct lighting: lighting by means of luminaires having a distribution of luminous intensity such that the fraction of the emitted luminous flux directly reaching the working plane, assumed to be unbounded, is 90–100 per cent.

Direct solar radiation: that part of the extraterrestrial solar radiation that as a collimated beam reaches the Earth's surface after selective attenuation by the atmosphere.

Directional lighting: lighting in which the light on the working plane or on an object is incident predominantly from a particular direction.

Disability glare: glare that impairs the vision of objects without necessarily causing discomfort.

Note: disability glare may be produced directly or by reflection.

For specification, disability glare may be expressed in a number of different ways. If threshold increment is used, the following values of TI should be used: 5%, 10%, 15%, 20%, 25%, 30%. If glare rating is used, then the following values of GR should be used: 10, 20, 30, 40, 45, 50, 55, 60, 70, 80, 90.

Discomfort glare: glare that causes discomfort without necessarily impairing the vision of objects.

Note: discomfort glare may be produced directly or by reflection.

For specification, if it is expressed using the unified glare rating the following values of UGR should be used: 10, 13, 16, 19, 21, 25, 28.

Downward light output ratio (of a luminaire): ratio of the downward flux of the luminaire, measured under specified practical conditions with its own lamps and equipment, to the sum of the individual luminous fluxes of the same lamps when operated outside the luminaire with the same equipment, under specified conditions.

Note: for luminaires using incandescent lamps only, the optical light output ratio and the light output ratio are the same in practice.

Emergency lighting: lighting provided for use when the supply to the normal lighting fails.

Flicker: impression of unsteadiness of visual sensation induced by a light stimulus whose luminance or spectral distribution fluctuates with time.

Floodlighting: lighting of a scene or object, usually by projectors, in order to increase considerably its illuminance relative to its surroundings.

Fusion frequency/critical flicker frequency (for a given set of conditions): frequency of alternation of stimuli above which flicker is not perceptible.

General colour-rendering index (of a light source) (R_a): value intended to specify the degree to which objects illuminated by a light source have an expected colour relative to their colour under a reference light source.

Note: R_a is derived from the colour rendering indices for a specified set of eight test colour samples. R_a has a maximum of 100, which generally occurs when the spectral distributions of the light source and the reference light source are substantially identical.

For specification, for design purposes colour rendering requirements should be specified using the general colour rendering index and should take one of the following values of R_a: 20, 40, 60, 80, 90.

General diffused lighting: lighting by means of luminaires having a distribution of luminous intensity such that the fraction of the emitted luminous flux directly reaching the working plane, assumed to be unbounded, is 40–60 per cent.

General lighting: substantially uniform lighting of an area without provision for special local requirements.

Glare: condition of vision in which there is discomfort or a reduction in the ability to see details or objects, caused by an unsuitable distribution or range of luminance, or to extreme contrasts. See also: **Disability glare** and **Discomfort glare**.

Global solar radiation: combined direct solar radiation and diffuse sky radiation.

Hemispherical illuminance (at a point) (E_{hs}): total luminous flux falling on the curved surface of a very small hemisphere located at the specified point divided by the curved surface area of the hemisphere. The base of the hemisphere is taken to be horizontal unless stated otherwise. Unit: lux.

Illuminance (at a point of a surface) (E): quotient of the luminous flux $d\Phi_v$, incident on an element of the surface containing the point, by the area dA of that element.

Equivalent definition: integral, taken over the hemisphere visible from the given point, of the expression $L \cos d\Omega$, where L is the luminance at the given point in the various directions of the incident elementary beams of solid angle $d\Omega$, and θ is the angle between any of these beams and the normal to the surface at the given point. Unit: lux (lx) = lumens per square metre.

Note: the orientation of the surface may be defined, e.g. horizontal, vertical; hence horizontal illuminance, vertical illuminance.

For specification, illuminance should be specified as maintained illuminance and should take one of the following values: 1.0×10^N lux, 1.5×10^N lux, 2.0×10^N lux, 3.0×10^N lux, 5.0×10^N lux, 7.5×10^N lux (where N is an integer). The area over which the illuminance is to be calculated or measured shall be specified.

Illuminance uniformity: ratio of minimum illuminance to average illuminance on a surface.

Note: use is also made of the ratio of minimum illuminance to maximum illuminance, in which case this should be specified explicitly.

Indirect lighting: lighting by means of luminaires having a distribution of luminous intensity such that the fraction of the emitted luminous flux directly reaching the working plane, assumed to be unbounded, is 0–10 per cent.

Initial illuminance: average illuminance when the installation is new. Unit: lux.

Initial luminance: (L_1): average luminance when the installation is new.

Note: the relevant points at which the luminances are determined shall be specified in the appropriate application standard.

Installed loading: the installed power of the lighting installation per unit area (for interior and exterior areas) or per unit length (for road lighting). Unit: watts per square metre (for areas), or kilowatts per kilometre (for road lighting).

Lamp: source made in order to produce an optical radiation, usually visible.

Note: this term is also sometimes used for certain types of luminaires.

Lamp lumen maintenance factor: ratio of the luminous flux of a lamp at a given time in its life to the initial luminous flux.

Lamp survival factor: fraction of the total number of lamps that continue to operate at a given time under defined conditions and switching frequency.

Life of lighting installation: period after which the installation cannot be restored to satisfy the required performance because of non-recoverable deterioration.

Light output ratio (of a luminaire): ratio of the total flux of the luminaire, measured under specified practical conditions with its own lamps and equipment, to the sum of the individual luminous fluxes of the same lamps when operated outside the luminaire with the same equipment, under specified conditions.

Note: for luminaires using incandescent lamps only, the optical light output ratio and the light output ratio are the same in practice.

Light output ratio working (of a luminaire) (η_w): ratio of the total flux of the luminaire, measured under specified practical conditions with its own lamps and equipment, to the sum of the individual luminous fluxes of the same lamps when operating outside the luminaire with a reference ballast, under reference conditions.

Local lighting: lighting for a specific visual task, additional to and controlled separately from the general lighting.

Localised lighting: lighting designed to illuminate an area with a higher illuminance at certain specified positions, for instance those at which work is carried out.

Luminaire: apparatus that distributes, filters or transforms the light transmitted from one or more lamps and which includes, except the lamps themselves, all the parts necessary for fixing and protecting the lamps and, where necessary, circuit auxiliaries together with the means for connecting them to the electric supply.

Note: the term 'lighting fitting' is deprecated.

Luminaire maintenance factor: ratio of the light output ratio of a luminaire at a given time to the initial light output ratio.

Luminance (L): luminous flux per unit solid angle transmitted by an elementary beam passing through the given point and propagating in the given direction, divided by the area of a section of that beam normal to the direction of the beam and containing the given point. It can also be defined as:

- the luminous intensity of the light emitted or reflected in a given direction from an element of the surface, divided by the area of the element projected in the same direction
- the illuminance produced by the beam of light on a surface normal to its direction, divided by the solid angle of the source as seen from the illuminated surface.

It is the physical measurement of the stimulus that produces the sensation of brightness. Unit: candela per square metre.

Technically defined, luminance is the quantity defined by the formula $L = d\Phi/(dA \cos\theta \, d\Omega)$, where $d\Phi$ is the luminous flux transmitted by an elementary beam passing through the given point and propagating in the solid angle $d\Omega$ containing the given direction; dA is the area of a section of that beam containing the given point; and θ is the angle between the normal to that section and the direction of the beam.

For specification, luminance should be specified as maintained luminance and should take one of the following values: 1.0×10^N cd/m^{-2}, 1.5×10^N cd/m^{-2}, 2.0×10^N cd/m^{-2}, 3.0×10^N cd/m^{-2}, 5.0×10^N cd/m^{-2}, 7.5×10^N cd/m^{-2} (where N is an integer). The area over which the luminance is to be calculated or measured shall be specified.

Luminance contrast: physical quantity intended to correlate with brightness contrast, usually defined by one of a number of formulae that involve the luminances of the stimuli considered (see also **Contrast**).

Note: luminance contrast may be defined as the luminance ratio, $C_1 = L_2/L_1$ (usually for successive contrasts), or by $C_2 = (L_2 - L_1)/L_1$ (usually for surfaces viewed simultaneously). When the areas of different luminance are comparable in size and it is desired to take an average, $C_3 = (L_2 - L_1)/0.5(L_2 + L_1)$ may be used instead, where L_1 is the luminance of the background (or largest part of the visual field) and L_2 is the luminance of the object.

Luminance meter: instrument for measuring luminance.

Luminance uniformity: ratio of minimum luminance to average luminance.

Note: use is also made of the ratio of minimum luminance to maximum luminance, in which case this should be specified explicitly.

Luminous efficacy of a source (η): quotient of the luminous flux emitted by the power consumed by the source. Unit: lumens per watt.

Notes: (1) it must be specified whether or not the power dissipated by auxiliary equipment (such as ballasts etc.), if any, is included in the power consumed by the source; (2) if not otherwise specified, the measurement conditions should be the reference conditions specified in the relevant IEC standard (see **Rated luminous flux**).

Luminous environment: lighting considered in relation to its physiological and psychological effects.

Luminous flux (Φ): quantity derived from radiant flux (radiant power) by evaluating the radiation according to the spectral sensitivity of the human eye (as defined by the CIE standard photometric observer). It is the light power emitted by a source or received by a surface. Unit : lumen (lm).

Notes: (1) in this definition, the values used for the spectral sensitivity of the CIE standard photometric observer are those of the spectral luminous efficiency function $V(\lambda)$; (2) see *IEC 50 (845); CIE 17.4; 845-01-22* for the definition of spectral luminous efficiency, *845-01-23* for the definition of the CIE standard photometric observer, and *845-01-56* for the definition of luminous efficacy of radiation. See also *ISO/CIE 10527*.

Technically defined as:

$$\Phi = K_{\mathrm{m}} \int_{-}^{\infty} (d\Phi_{\mathrm{e}}(\lambda)/d\lambda)\nu(\lambda)d(\lambda)$$

where $d\Phi(\lambda)/d\lambda$ is the spectral distribution of radiant flux, and $\nu(\lambda)$ is the spectral luminous efficiency.

Luminous intensity (of a point source in a given direction) (I): luminous flux per unit solid angle in the direction in question, i.e. the luminous flux on a small surface, divided by the solid angle that the surface subtends at the source. Unit: candela = lumen per steradian.

Technically defined as quotient of the luminous flux $d\Phi$ leaving the source and propagated in the element solid angle $d\Omega$ containing the given direction, by the element solid angle. $I = d\Phi/d\Omega$.

Maintained illuminance: value below which the average illuminance on the specified area should not fall. It is the average illuminance at the time maintenance should be carried out. Unit: lux.

Maintenance cycle: repetition of lamp replacement, lamp/luminaire cleaning and room surface cleaning intervals.

Maintenance factor: ratio of the average illuminance on the working plane after a certain period of use of a lighting installation to the average illuminance obtained under the same condition for the installation considered conventionally as new.

Notes: (1) the term depreciation factor has been formerly used to designate the reciprocal of the above ratio; (2) the maintenance factor of an installation depends on lamp lumen maintenance factor, lamp survival factor, luminaire maintenance factor and (for an interior lighting installation) room surface maintenance factor.

Maintenance schedule: set of instructions specifying maintenance cycle and servicing procedures.

Maximum illuminance: highest illuminance at any relevant point on the specified surface. Unit: lux.

Note: the relevant points at which the illuminances are determined shall be specified in the appropriate application standard.

Maximum luminance (L_{max}): highest luminance of any relevant point on the specified surface. Unit: candela per square metre.

Note: the relevant points at which the luminances are determined shall be specified in the appropriate application standard.

Measurement field (of a photometer): area including all points in object space, radiating towards the acceptance area of the detector.

Minimum illuminance: lowest illuminance at any relevant point on the specified surface. Unit: lux.

Note: the relevant points at which the illuminances are determined shall be specified in the appropriate application standard.

Minimum luminance (L_{min}): lowest luminance of any relevant point on the specified surface. Unit: candela per square metre.

Note: the relevant points at which the luminances are determined shall be specified in the appropriate application standard.

Photometer: instrument for measuring photometric quantities.

Photometry: measurement of quantities referring to radiation evaluated according to the sensitivity of the human eye (as defined by the CIE standard photometric observer).

Notes: (1) the values usually used for the spectral sensitivity of the CIE standard photometric observer are those of the spectral luminous efficiency function $V(\lambda)$; (2) see **Luminous flux** for the definition of spectral luminous efficiency.

Rated luminous flux (of a type of lamp): the value of the initial luminous flux of a given type of lamp declared by the manufacturer or the responsible vendor, the lamp being operated under specified conditions. Unit: lumens.

Notes: (1) for most lamps, in reference conditions the lamps is usually operating at am ambient temperature of 25°C in air, freely suspended in a defined burning position and with a reference ballast, but see the relevant IEC standard for the particular lamp; (2) the initial luminous flux is the luminous flux of a lamp after a short ageing period as specified in the relevant lamp standard; (3) the rated luminous flux is sometimes marked on the lamp.

Reference ballast: a special inductive-type ballast designed for the purpose of providing comparison standards for use in testing ballasts, for the selection of reference lamps and for testing regular production lamps under standardised conditions.

Reference lamp: a discharge lamp selected for the purpose of testing ballasts and which, when associated with a reference ballast under specified conditions, has electrical values that are close to the objective values given in a relevant specification.

Reference surface: surface on which illuminance is measured or specified.

Reflectance (ρ): ratio of luminous flux reflected from a surface to the luminous flux incident on it.

Note: the reflectance generally depends on the spectral distribution and polarization of the incident light, the surface finish, and the geometry of the incident and reflected light relative to the surface.

Reflectometer: instrument for measuring quantities pertaining to reflection.

Room surface maintenance factor: ratio of room surface reflectance at a given time to the initial reflectance value.

Semi-cylindrical illuminance (at a point) (E_{SZ}): total luminous flux falling on the curved surface of a very small semi-cylinder located at the specified point, divided by the curved surface area of the semi-cylinder. The axis of the semi-cylinder is taken to be vertical unless stated otherwise. The direction of the curved surface should be specified. Unit: lux.

Semi-direct lighting: lighting by means of luminaires having a distribution of luminous intensity such that the fraction of the emitted luminous flux directly reaching the working plane, assumed to be unbounded, is 60–90 per cent.

Semi-indirect lighting: lighting by means of luminaires having a distribution of luminous intensity such that the fraction of the emitted luminous flux directly reaching the working plane, assumed to be unbounded, is 10–40 per cent.

Solar radiation: electromagnetic radiation from the sun.

Spacing (in an installation): distance between the light centres of adjacent luminaires of the installation.

Spacing-to-height ratio: ratio of spacing to the height of the geometric centres of the luminaires above the reference plane.

Note: for indoor lighting the reference plane is usually the horizontal working plane; for exterior lighting the reference plane is usually the ground.

(Spatial) distribution of luminous intensity (of a lamp or luminaire): display, by means of curves or tables, of the value of the luminous intensity of the source as a function of direction in space.

Spherical illuminance (at a point) (E_o): total luminous flux falling onto the whole surface of very small sphere located at the specified point divided by the total surface area of the sphere. Unit: lux.

Technically defined by the formula:

$$E_o = \int_{4\pi sr} L d\Omega$$

where $d\Omega$ is the solid angle of each elementary beam passing through the given point and L is its luminance at that point.

Spotlighting: lighting designed to increase considerably the illuminance of a limited area or of an object relative to the surroundings, with minimum diffused lighting.

Stroboscopic effect: apparent change of motion and/or appearance of a moving object when the object is illuminated by a light of varying intensity.

Note: to obtain apparent immobilisation or constant change of movement, it is necessary that both the object movement and the light intensity variation are periodic, and that some specific relation between the object movement and light variation frequencies exists. The effect is only observable if the amplitude of the light variation is above certain limits. The motion of the object may be rotational or translational.

Sunlight: visible part of direct solar radiation.

Note: when dealing with actinic effects of optical radiations, this term is commonly used for radiations extending beyond the visible region of the spectrum.

Transmittance (τ): Ratio of the luminous flux transmitted through a body to the luminous flux incident on it.

Note: the transmittance generally depends on the direction, polarization and spectral distribution of the incident light and on the surface finish.

Tristimulus values (of a colour stimulus): amounts of the three reference colour stimuli, in a given trichromatic system, required to match the colour of the stimulus considered.

Note: in the CIE standard colorimetric systems, the tristimulus values are represented by the symbols X, Y, Z and X_{10}, Y_{10}, Z_{10}.

Unified glare rating: see **Disability glare**.

Upward light output ratio (of a luminaire): ratio of the upward flux of the luminaire, measured under specified practical conditions with its own lamps and equipment, to the sum of the individual luminous fluxes of the same lamps when operated outside the luminaire with the same equipment, under specified conditions.

Note: for luminaires using incandescent lamps only, the optical light output ratio and the light output ratio are the same in practice.

Utilisation factor: ratio of the luminous flux received by the reference surface to the sum of the rated lamp luminous fluxes of the lamps in the installation.

V(λ) correction: correction of the spectral responsivity of a detector to match the photopic spectral sensitivity of the human eye.

Veiling reflections: specular reflections that appear on the object viewed, and partially or wholly obscure details by reducing contrast.

Visual acuity: capacity for seeing distinctly fine details that have a very small angular subtense at the eye.

Note: quantitatively, visual acuity can be expressed by the reciprocal of the angle, in minutes of arc, subtended at the entrance pupil by the extremities of the detail separation that is just visible.

Technically defined, (1) qualitatively, as the capacity for seeing distinctly fine details that have very small angular separation; (2) quantitatively, as any of a number of measures of spatial discrimination such as the reciprocal of the value of the angular separation in minutes of arc of two neighbouring objects (points or lines or other specified stimuli) that the observer can just perceive to be separate.

Visual comfort: subjective condition of visual well being induced by the visual environment.

Visual field: area or extent of physical space visible to an eye at a given position and direction of view.
Note: the visual field may be either monocular or binocular.

Visual performance: performance of the visual system as measured, for instance, by the speed and accuracy with which a visual task is performed.

Index